Synthesis, Characterization and Performance Enhancement of Electrode and Biomaterial Coatings

Synthesis, Characterization and Performance Enhancement of Electrode and Biomaterial Coatings

Editors

Shang Wang
Qing Sun
Chenyu Liu

Basel • Beijing • Wuhan • Barcelona • Belgrade • Novi Sad • Cluj • Manchester

Editors

Shang Wang
Harbin Institute of Technology
Harbin
China

Qing Sun
Harbin Institute of Technology
Harbin
China

Chenyu Liu
Dalian University of Technology
Dalian
China

Editorial Office
MDPI AG
Grosspeteranlage 5
4052 Basel, Switzerland

This is a reprint of articles from the Special Issue published online in the open access journal *Coatings* (ISSN 2079-6412) (available at: https://www.mdpi.com/journal/coatings/special_issues/Electrode_Biomaterial_Coatings).

For citation purposes, cite each article independently as indicated on the article page online and as indicated below:

Lastname, A.A.; Lastname, B.B. Article Title. *Journal Name* **Year**, *Volume Number*, Page Range.

ISBN 978-3-7258-2597-4 (Hbk)
ISBN 978-3-7258-2598-1 (PDF)
doi.org/10.3390/books978-3-7258-2598-1

© 2024 by the authors. Articles in this book are Open Access and distributed under the Creative Commons Attribution (CC BY) license. The book as a whole is distributed by MDPI under the terms and conditions of the Creative Commons Attribution-NonCommercial-NoDerivs (CC BY-NC-ND) license.

Contents

About the Editors vii

Chenyu Liu, Qing Sun and Shang Wang
Special Issue "Synthesis, Characterization and Performance Enhancement of Electrode and Biomaterial Coatings"
Reprinted from: *Coatings* 2024, 14, 1276, doi:10.3390/coatings14101276 1

Geng Li, Shang Wang, Jiayue Wen, Shujun Wang, Yuxin Sun, Jiayun Feng and Yanhong Tian
A Manufacturing Method for High-Reliability Multilayer Flexible Electronics by Electrohydrodynamic Printing
Reprinted from: *Coatings* 2024, 14, 625, doi:10.3390/coatings14050625 4

Wenyu Wu, Geng Li, Shang Wang, Yiping Wang, Jiayun Feng, Xiaowei Sun and Yanhong Tian
Study on the Solder Joint Reliability of New Diamond Chip Resistors for Power Devices
Reprinted from: *Coatings* 2023, 13, 748, doi:10.3390/coatings13040748 14

Michael M. Slepchenkov, Pavel V. Barkov and Olga E. Glukhova
Electronic and Electrical Properties of Island-Type Hybrid Structures Based on Bi-Layer Graphene and Chiral Nanotubes: Predictive Analysis by Quantum Simulation Methods
Reprinted from: *Coatings* 2023, 13, 966, doi:10.3390/coatings13050966 25

Ruiqin Peng, Xuzhen Zhuang, Yuanyuan Li, Zhiguo Yu and Lijie Ci
High Gas Response Performance Based on Reduced Graphene Oxide/SnO_2 Nanowires Heterostructure for Triethylamine Detection
Reprinted from: *Coatings* 2023, 13, 849, doi:10.3390/coatings13050849 43

Yucheng Zhu, Xiaofei Cao, Yuan Tan, Yao Wang, Jun Hu, Baotong Li and Zhong Chen
Single-Layer MoS_2: A Two-Dimensional Material with Negative Poisson's Ratio
Reprinted from: *Coatings* 2023, 13, 283, doi:10.3390/coatings13020283 52

Yun Chen, Bo Jiang, Yue Zhao, Hongbin Liu and Tingli Ma
Diatomite and Glucose Bioresources Jointly Synthesizing Anode/Cathode Materials for Lithium-Ion Batteries
Reprinted from: *Coatings* 2023, 13, 146, doi:10.3390/coatings13010146 63

Elena Olegovna Nasakina, Maria Andreevna Sudarchikova, Konstantin Yurievich Demin, Alexandra Borisovna Mikhailova, Konstantin Vladimirovich Sergienko, Sergey Viktorovich Konushkin, et al.
Study of Co-Deposition of Tantalum and Titanium during the Formation of Layered Composite Materials by Magnetron Sputtering
Reprinted from: *Coatings* 2023, 13, 114, doi:10.3390/coatings13010114 73

Ili Salwani Mohamad, Mohd Natashah Norizan, Norsuria Mahmed, Nurnaeimah Jamalullail, Dewi Suriyani Che Halin, Mohd Arif Anuar Mohd Salleh, et al.
Enhancement of Power Conversion Efficiency with Zinc Oxide as Photoanode and *Cyanococcus*, *Punica granatum* L., and *Vitis vinifera* as Natural Fruit Dyes for Dye-Sensitized Solar Cells
Reprinted from: *Coatings* 2022, 12, 1781, doi:10.3390/coatings12111781 91

Farishta Shafiq, Simiao Yu, Yongxin Pan and Weihong Qiao
Synthesis and Characterization of Titania-Coated Hollow Mesoporous Hydroxyapatite Composites for Photocatalytic Degradation of Methyl Red Dye in Water
Reprinted from: *Coatings* 2024, 14, 921, doi:10.3390/coatings14080921 104

Ru Yang, Yongfa Diao, Hongbin Liu and Yihang Lu
Experimental and Adsorption Kinetics Study of Hg^0 Removal from Flue Gas by Silver-Loaded Rice Husk Gasification Char
Reprinted from: *Coatings* **2024**, *14*, 797, doi:10.3390/coatings14070797 . . . . **125**

About the Editors

Shang Wang

Shang Wang earned his B.S. and Ph.D. from the Harbin Institute of Technology in 2014 and 2019, respectively. He has served as a Lecturer and an Associate Professor, and he currently holds the position of Deputy Director of the Department of Welding Science and Engineering. Dr. Wang has led numerous national and provincial research projects, including two funded by the National Natural Science Foundation of China. He has published over 90 peer-reviewed papers and holds several invention patents. His research interests include advanced electronic packaging technology and flexible electronic device applications.

Qing Sun

Qing Sun obtained his B.S. from the Harbin Institute of Technology (HIT) in 2014 and his Ph.D. from Shandong University in 2022. He subsequently conducted research at the Shenzhen campus of the HIT and the Catalonia Institute for Energy Research. He is now working as an Associate Professor at the Zhengzhou Research Institute of HIT. He has led and participated in several research projects funded by the EU and national, and provincial governments. He has published over 50 peer-reviewed papers, which have been cited more than 2,400 times, with an H-index of 26. His research focuses on the development of energy storage materials and flexible energy storage devices.

Chenyu Liu

Chenyu Liu earned her B.S. and Ph.D. from the Dalian University of Technology in 2014 and 2019, respectively, and was a joint doctoral student at the University of California, Santa Barbara, from 2015 to 2017. She has served as an Associate Professor at the State Key Laboratory of Fine Chemicals at the Dalian University of Technology. Her research focuses on the structure–activity relationships and mechanisms of amphiphilic compounds, particularly in the areas of biomedical amphiphilic materials, oil recovery systems, and functional surfactants. Dr. Liu has received several accolades, including the Eighth Young Elite Scientists Sponsorship Program by CAST and the Dalian Science and Technology Star award. She has led numerous national and provincial research projects, including two funded by the National Natural Science Foundation of China. To date, she has published over 50 peer-reviewed papers and holds several invention patents.

Editorial

Special Issue "Synthesis, Characterization and Performance Enhancement of Electrode and Biomaterial Coatings"

Chenyu Liu [1], Qing Sun [2] and Shang Wang [2,*]

[1] State Key Laboratory of Fine Chemicals, School of Chemical Engineering, Dalian University of Technology, Dalian 116024, China
[2] State Key Laboratory of Precision Welding & Joining of Materials and Structures, Harbin Institute of Technology, Harbin 150001, China
* Correspondence: wangshang@hit.edu.cn

Citation: Liu, C.; Sun, Q.; Wang, S. Special Issue "Synthesis, Characterization and Performance Enhancement of Electrode and Biomaterial Coatings". *Coatings* **2024**, *14*, 1276. https://doi.org/10.3390/coatings14101276

Received: 24 September 2024
Accepted: 29 September 2024
Published: 7 October 2024

Copyright: © 2024 by the authors. Licensee MDPI, Basel, Switzerland. This article is an open access article distributed under the terms and conditions of the Creative Commons Attribution (CC BY) license (https://creativecommons.org/licenses/by/4.0/).

Functional materials are extensively employed across diverse domains, including energy storage systems [1,2], electronic devices [3], and medical implants [4], where their performance significantly influences the efficiency, reliability, and longevity of related technologies [5,6]. However, in real-world applications, these materials often encounter substantial challenges [7], such as inadequate electrical conductivity, limited mechanical strength, severe interfacial side reactions in electrode materials, and poor biocompatibility and corrosion resistance in biomaterials. To address these issues, surface coating modification has emerged as a highly effective strategy to enhance their physical, chemical, and mechanical properties [8], thereby meeting the multifaceted demands of modern applications.

Surface coating technologies for functional materials have garnered significant attention within the materials science and engineering community [9], particularly in improving the performance of critical components such as electrodes and biomaterials. For example, electrode materials in energy storage systems—such as lithium-ion batteries (LIBs) and electrochemical capacitors—are prone to mechanical degradation and undesirable side reactions due to repetitive charge/discharge cycles, which can lead to substantial performance deterioration over time [10]. Coating these electrodes with materials exhibiting superior electrical conductivity and chemical stability, including metal oxides, ceramics, graphene, and conductive polymers, can markedly enhance their structural integrity [11]. This, in turn, optimizes the interfacial reaction kinetics between the electrode and electrolyte [12], resulting in improved cycle stability and enhanced capacity retention [13]. Moreover, surface coatings can significantly lower the intrinsic resistance of electrode materials and boost ion transport rates [14], both pivotal factors for enhancing the power density and operational efficiency of energy storage devices.

Coating technologies also demonstrate remarkable potential in advancing biomaterials [15]. For medical implants, biocompatibility, corrosion resistance, and controlled degradability are paramount for ensuring seamless integration with human tissues and long-term safety. Metals, commonly used in orthopedic and dental implants [16], often exhibit limited corrosion resistance and biocompatibility, potentially triggering adverse biological reactions. By applying bioactive coatings such as bioceramics or biofunctional polymers to metal surfaces, these limitations can be mitigated, leading to improved tissue integration and promoting the regeneration of bone or other tissues. For instance, bioceramic coatings such as hydroxyapatite [17], which closely resemble the inorganic composition of bone, facilitate faster osseointegration, while conductive polymer coatings can enhance the electrical properties of implants, proving especially advantageous in applications such as neural tissue engineering.

Notably, surface coatings not only augment the intrinsic properties of functional materials but also allow for precise optimization of targeted functionalities by carefully controlling parameters such as coating thickness, microstructure, and composition. For example, ceramic coatings are widely adopted in high-temperature electrode materials due to their excellent thermal stability and corrosion resistance, whereas metal oxide coatings are frequently employed in electrochemical storage devices for their robust chemical inertness. Additionally, graphene and two-dimensional materials [18,19], known for their exceptional conductivity and mechanical flexibility as coatings, are optimal for improving the electrochemical performance of LIBs and supercapacitors. By judiciously selecting and designing coating materials, it is possible to enhance conductivity, reinforce mechanical durability, minimize detrimental side reactions, and elevate the overall performance of energy storage systems.

This Special Issue seeks to provide an in-depth exploration of the design, fabrication, and application of surface coatings for functional materials. Emphasis will be placed on the comparative evaluation and optimization of various coating technologies to elucidate the underlying mechanisms by which surface modification strategies improve material performance. The scope includes the investigation of ceramics, metals, metal oxides, graphene, and conductive polymers, both as individual materials and as surface coatings, with a focus on their roles in enhancing mechanical robustness and electronic and ionic conductivity and mitigating interfacial degradation. Furthermore, the Issue will address emerging trends and future directions in coating technology, offering novel insights and guidelines for the development of next-generation coatings for functional materials across a wide range of applications.

Author Contributions: Conceptualization, C.L.; writing—original draft preparation, C.L. and Q.S.; writing—review and editing, S.W. All authors have read and agreed to the published version of the manuscript.

Funding: S.W. acknowledges the financial support provided by the National Natural Science Foundation of China (52105329).

Acknowledgments: C.L., Q.S., and S.W. thank all of the authors for their contributions to this Special Issue, entitled "Synthesis, Characterization and Performance Enhancement of Electrode and Biomaterial Coatings", and the editorial staff of the journal *Coatings*.

Conflicts of Interest: The authors declare no conflicts of interest.

References

1. Mohamad, I.S.; Norizan, M.N.; Mahmed, N.; Jamalullail, N.; Halin, D.S.C.; Salleh, M.A.A.M.; Sandu, A.V.; Baltatu, M.S.; Vizureanu, P. Enhancement of Power Conversion Efficiency with Zinc Oxide as Photoanode and Cyanococcus, Punica granatum L., and Vitis vinifera as Natural Fruit Dyes for Dye-Sensitized Solar Cells. *Coatings* **2022**, *12*, 1781. [CrossRef]
2. Sun, Q.; Li, J.; Yang, M.; Wang, S.; Zeng, G.; Liu, H.; Cheng, J.; Li, D.; Wei, Y.; Si, P.; et al. Carbon Microstructure Dependent Li-Ion Storage Behaviors in SiOx/C Anodes. *Small* **2023**, *19*, 2300759. [CrossRef] [PubMed]
3. Wu, W.; Li, G.; Wang, S.; Wang, Y.; Feng, J.; Sun, X.; Tian, Y. Study on the Solder Joint Reliability of New Diamond Chip Resistors for Power Devices. *Coatings* **2023**, *13*, 748. [CrossRef]
4. Yang, C.; Luo, Y.; Shen, H.; Ge, M.; Tang, J.; Wang, Q.; Lin, H.; Shi, J.; Zhang, X. Inorganic nanosheets facilitate humoral immunity against medical implant infections by modulating immune co-stimulatory pathways. *Nat. Commun.* **2022**, *13*, 4866. [CrossRef] [PubMed]
5. Yang, R.; Diao, Y.; Liu, H.; Lu, Y. Experimental and Adsorption Kinetics Study of Hg0 Removal from Flue Gas by Silver-Loaded Rice Husk Gasification Char. *Coatings* **2024**, *14*, 797. [CrossRef]
6. Li, G.; Wang, S.; Wen, J.; Wang, S.; Sun, Y.; Feng, J.; Tian, Y. A Manufacturing Method for High-Reliability Multilayer Flexible Electronics by Electrohydrodynamic Printing. *Coatings* **2024**, *14*, 625. [CrossRef]
7. Zeng, G.; Sun, Q.; Horta, S.; Wang, S.; Lu, X.; Zhang, C.Y.; Li, J.; Li, J.; Ci, L.; Tian, Y.; et al. A Layered Bi_2Te_3@PPy Cathode for Aqueous Zinc-Ion Batteries: Mechanism and Application in Printed Flexible Batteries. *Adv. Mater.* **2024**, *36*, 2305128. [CrossRef] [PubMed]
8. Sun, Q.; Yang, M.; Zeng, G.; Li, J.; Hu, Z.; Li, D.; Wang, S.; Si, P.; Tian, Y.; Ci, L. Insights into the Potassium Ion Storage Behavior and Phase Evolution of a Tailored Yolk–Shell SnSe@C Anode. *Small* **2022**, *18*, 2203459. [CrossRef] [PubMed]
9. Nasakina, E.O.; Sudarchikova, M.A.; Demin, K.Y.; Mikhailova, A.B.; Sergienko, K.V.; Konushkin, S.V.; Kaplan, M.A.; Baikin, A.S.; Sevostyanov, M.A.; Kolmakov, A.G. Study of Co-Deposition of Tantalum and Titanium during the Formation of Layered Composite Materials by Magnetron Sputtering. *Coatings* **2023**, *13*, 114. [CrossRef]
10. Sun, Q.; Zeng, G.; Li, J.; Wang, S.; Botifoll, M.; Wang, H.; Li, D.; Ji, F.; Cheng, J.; Shao, H.; et al. Is Soft Carbon a More Suitable Match for SiOx in Li-Ion Battery Anodes? *Small* **2023**, *19*, 2302644. [CrossRef] [PubMed]

11. Chen, Y.; Jiang, B.; Zhao, Y.; Liu, H.; Ma, T. Diatomite and Glucose Bioresources Jointly Synthesizing Anode/Cathode Materials for Lithium-Ion Batteries. *Coatings* 2023, *13*, 146. [CrossRef]
12. Wang, S.; Zeng, G.; Sun, Q.; Feng, Y.; Wang, X.; Ma, X.; Li, J.; Zhang, H.; Wen, J.; Feng, J.; et al. Flexible Electronic Systems via Electrohydrodynamic Jet Printing: A MnSe@rGO Cathode for Aqueous Zinc-Ion Batteries. *ACS Nano* 2023, *17*, 13256–13268. [CrossRef] [PubMed]
13. Peng, R.; Zhuang, X.; Li, Y.; Yu, Z.; Ci, L. High Gas Response Performance Based on Reduced Graphene Oxide/SnO_2 Nanowires Heterostructure for Triethylamine Detection. *Coatings* 2023, *13*, 849. [CrossRef]
14. Sun, Q.; Li, D.; Dai, L.; Liang, Z.; Ci, L. Structural Engineering of SnS_2 Encapsulated in Carbon Nanoboxes for High-Performance Sodium/Potassium-Ion Batteries Anodes. *Small* 2020, *16*, 2005023. [CrossRef] [PubMed]
15. Lee, S.; Lee, J.; Byun, H.; Kim, S.-j.; Joo, J.; Park, H.H.; Shin, H. Evaluation of the anti-oxidative and ROS scavenging properties of biomaterials coated with epigallocatechin gallate for tissue engineering. *Acta Biomater.* 2021, *124*, 166–178. [CrossRef] [PubMed]
16. Al-Shalawi, F.D.; Mohamed Ariff, A.H.; Jung, D.-W.; Mohd Ariffin, M.K.A.; Seng Kim, C.L.; Brabazon, D.; Al-Osaimi, M.O. Biomaterials as Implants in the Orthopedic Field for Regenerative Medicine: Metal versus Synthetic Polymers. *Polymers* 2023, *15*, 2601. [CrossRef] [PubMed]
17. Shafiq, F.; Yu, S.; Pan, Y.; Qiao, W. Synthesis and Characterization of Titania-Coated Hollow Mesoporous Hydroxyapatite Composites for Photocatalytic Degradation of Methyl Red Dye in Water. *Coatings* 2024, *14*, 921. [CrossRef]
18. Slepchenkov, M.M.; Barkov, P.V.; Glukhova, O.E. Electronic and Electrical Properties of Island-Type Hybrid Structures Based on Bi-Layer Graphene and Chiral Nanotubes: Predictive Analysis by Quantum Simulation Methods. *Coatings* 2023, *13*, 966. [CrossRef]
19. Zhu, Y.; Cao, X.; Tan, Y.; Wang, Y.; Hu, J.; Li, B.; Chen, Z. Single-Layer MoS_2: A Two-Dimensional Material with Negative Poisson's Ratio. *Coatings* 2023, *13*, 283. [CrossRef]

Disclaimer/Publisher's Note: The statements, opinions and data contained in all publications are solely those of the individual author(s) and contributor(s) and not of MDPI and/or the editor(s). MDPI and/or the editor(s) disclaim responsibility for any injury to people or property resulting from any ideas, methods, instructions or products referred to in the content.

Article

A Manufacturing Method for High-Reliability Multilayer Flexible Electronics by Electrohydrodynamic Printing

Geng Li [1,2], Shang Wang [1,2,3,*], Jiayue Wen [1,2,*], Shujun Wang [1], Yuxin Sun [1], Jiayun Feng [1] and Yanhong Tian [1,2]

1. National Key Laboratory of Precision Welding & Joining of Materials and Structures, Harbin Institute of Technology, Harbin 150001, China; li-geng@hit.edu.cn (G.L.); shujunwang0103@163.com (S.W.); sunyuxin_2004@163.com (Y.S.); fengjy@hit.edu.cn (J.F.); tianyh@hit.edu.cn (Y.T.)
2. Zhengzhou Research Institute, Harbin Institute of Technology, Zhengzhou 450000, China
3. State Key Laboratory of Precision Electronic Manufacturing Technology and Equipment, Guangdong University of Technology, Guangzhou 510006, China
* Correspondence: wangshang@hit.edu.cn (S.W.); wenjiayue_ep@hit.edu.cn (J.W.)

Abstract: To meet the demand for higher performance and wearability, integrated circuits are developing towards having multilayered structures and greater flexibility. However, traditional circuit fabrication methods using etching and lamination processes are not compatible with flexible substrates. As a non-contact printing method in additive manufacturing, electrohydrodynamic printing possesses advantages such as environmental friendliness, sub-micron manufacturing, and the capability for flexible substrates. However, the interconnection and insulation of different conductive layers become significant challenges. This study took composite silver ink as a conductive material to fabricate a circuit via electrohydrodynamic printing, applied polyimide spraying to achieve interlayer insulation, and drilled micro through-holes to achieve interlayer interconnection. A 200×200 mm^2 ten-layer flexible circuit was thus prepared. Furthermore, we combined a finite element simulation with reliability experiments, and the prepared ten-layer circuit was found to have excellent bending resistance and thermal cycling stability. This study provides a new method for the manufacturing of low-cost, large-sized, multilayer flexible circuits, which can improve circuit performance and boost the development of printed electronics.

Keywords: multilayer circuit; additive manufacturing; printing electronics; reliability

1. Introduction

With the end of Moore's Law, integrated circuits have shifted from blindly pursuing an increase in integration to enhancing intelligence [1]. Intelligent electronics, including smart wearable devices, energy storage devices, sensors, and smart fabrics, are typically adhered to structures or biological surfaces to obtain more accurate sensing signals and enhanced wear comfort, which demands greater flexibility from these circuits [2–4]. Additionally, the superb conformal ability of flexible circuits allows them to adhere to the curved surfaces of three-dimensional objects without being limited by installation space. For example, conformal circuits can be applied to radar antennas on aircraft surfaces, structural health monitoring sensors, etc., significantly reducing equipment weight and enhancing aerodynamic performance [5–8]. However, traditional silicon-based circuit manufacturing methods, which utilize photolithography, etching, and other processes, are unsuitable for fabrication flexible substrates such as polyethylene terephthalate (PET) and polydimethylsiloxane (PDMS). Screen-printing technology is able to manufacture large-area conductive patterns on flexible substrates and is compatible with high-viscosity pastes, achieving superior conductivity. However, screen printing can only achieve a pattern resolution of at least a few hundred micrometers, thereby posing challenges for the fabrication of high-precision flexible circuits [9,10].

The advantages of inkjet printing as an additive manufacturing method are its high precision and environmental friendliness, and it is suitable for various substrates such as metals and flexible polymers [11–13]. Among the many inkjet printing methods, electrohydrodynamic (EHD) printing is compatible with various types of nanomaterials, and possesses the ability to prepare flexible electronics with an ultra-fine accuracy below 5 μm [14,15]. The principle of EHD printing is to apply a voltage between the nozzle and substrate, and the ink forms a Taylor cone under the action of the electric field that comes into contact with the substrate. At this time, the pathway conducts, the Taylor cone disappears, and then the circuit disconnects to generate a new Taylor cone [16]. During this cycle, the ink accumulates on the substrate with an ultra-fine resolution, and combined with a precision displacement platform, micro-scale patterns can be printed. Some researchers have prepared flexible heaters, supercapacitors, strain sensors, and other devices through EHD printing [14,17]. However, these circuits are almost all single-layer structures, which lead to larger volumes and weights compared to multilayer circuits with the same performance [18]. Multilayer circuits undoubtedly have more complex preparation processes; researchers use laser etching or photolithography to prepare flexible multilayer circuits, but these methods have high costs [19–21]. Therefore, it is necessary to develop a new multilayer circuit manufacturing method to achieve interlayer insulation and interconnection between multilayer circuits.

Bending stress is the most common working load in flexible electronics, which can lead to the cracking, delamination, and degradation of functional materials. Therefore, bending cycle tests are the most important indicator for evaluating the reliability of flexible electronics [22,23]. Especially for large multilayer circuits, their higher thickness leads to higher tensile/compressive stress at regions far from the neutral surface during bending, thereby increasing the risk of failure [24]. For circuits prepared via inkjet printing, it is difficult to ensure sufficient adhesion between the conductive layer and the insulation layer, and delamination may occur during the bending cycle process, which poses challenges to the selection of conductive ink and insulation layer materials. Given this research background, this study adopts a new flexible circuit fabrication method based on EHD printing, aiming to prepare large-area flexible multilayer electronics with excellent bending reliability. In order to make multilayer circuits more compatible with a wider range of usage scenarios, thermal cycling reliability is also considered.

2. Materials and Methods

2.1. Laboratory Experiment

The multilayer circuit manufacturing method developed in this study is shown in Figure 1. Firstly, the conductive path was printed onto a PET substrate via EHD, as shown in Figure 1a. The ink used for printing was WIK-36A high-conductivity composite silver ink from LOCTITE (Brussels, Belgium). Interlayer insulation was achieved by spraying insulated ink, as shown in Figure 1b. The insulated ink was obtained by mixing 5 mL of 20% polyimide (PI) solution with 15 mL of dimethylacetamide (DMAC) solution and stirring at 500 rpm for 0.5 h. The distance between the spray pen and the substrate was 20 mm, the spraying pressure was 0.1 MPa, and the time was 10 s per layer. After spraying the insulation layer, the PET substrate was transferred to a hot plate and dried at 80 °C for 5 min. Then, the initial operation was repeated for the next layer of conductive path printing. The method of interconnection between layers is shown in Figure 1c. We used a micro drilling bit to drill through-holes in the interconnection area and injected high-viscosity conductive paste into them. We designed ten-layer circuit patterns as shown in Figure 1d to verify the conductivity between each layer. Unconnected patterns were printed on the bottom layer, as shown in Figure S1 of the Supplementary Materials. We printed the square areas to serve as pads to drill through-holes for interlayer interconnection, and the circular areas served as pins to test connection status. By designing different conductive layer patterns, as shown in Figure S2, specific pins were connected, such as pin 1 and pin 2. When testing the resistance between different pins, the unconnected pins should be open circuit, while the connected pins are conductive. Therefore, whether effective interlayer interconnection has been achieved can be determined by the open circuit or conductivity between the pins.

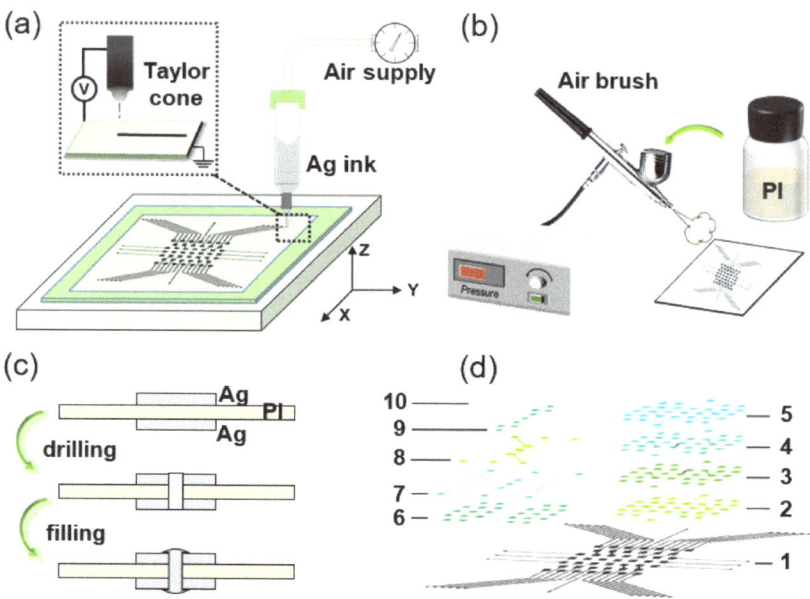

Figure 1. Schematic manufacturing process of multilayer flexible circuits: (**a**) electrohydrodynamic printing; (**b**) insulation; (**c**) interconnection; (**d**) conductive patterns of the 10-layer circuit.

In order to explore the process window of EHD printing, we conducted experiments at different printing speeds, nozzle heights, and nozzle diameters to determine the influence of various process parameters on printing quality. During the experiment, the electric field strength remained at 1.2×10^7 V/m. Furthermore, we investigated the sintering mechanism of conductive ink through post-treatment with heat.

We conducted bend cycling tests by using a self-developed testing platform with a bending angle of 120° and one bending cycle for 5 s. The thermal cycling test was conducted using a thermal cycling test chamber for a total of 1000 cycles. The thermal cycling profile was in accordance with the requirements of JESD22-A104: the temperature ranged from −55 °C to 85 °C, one cycle was held for 3000 s, with a rising time and falling time of 900 s and a cycle of 600 s at a consistent temperature. Five samples were prepared under each condition to exclude randomness. We measured the resistance between pin 1 and pin 2 (Figure S1) to evaluate the reliability of the multilayer circuit, because this conductive path was most affected by bending stress and passed through the most through-holes. Resistance and resistivity measurements were taken using four MCP-t370 probes by Mitsubishi Chemical (Tokyo, Japan). The microstructure and cross-section morphology of the multilayer flexible circuits were investigated using a scanning electron microscope (SEM, TESCAN CLARA, Brno, Czech), and the elemental composition was identified via energy dispersive spectrometry (EDS, TESCAN CLARA).

2.2. Numerical Simulation

In order to investigate the stress–strain distribution of multilayer circuits with different loads under thermal cycling and bend cycling, we conducted corresponding finite element analysis (FEA) simulation using ANSYS APDL (version 18.0). Several assumptions were made to ensure the accuracy and feasibility of the numerical simulation, as listed below:

- All materials were uniform and dense;
- All interconnecting interfaces were tightly combined;
- Thickness was constant in the same layer;
- Changes in a material's thermodynamic parameters with temperature were considered in the thermal cycling simulation.

The numerical simulation process is shown in Figure 2. The FEA model was established by APDL commands as shown in Figure 2a. Thermodynamic simulation used thermal element SOLID70 and structural element SOLID185. Tetrahedral elements were used for meshing, and the elements around the conductive parts were refined to improve the simulation accuracy, as shown in Figure 2b. The insulation layer between the conductive layers is very thin (within 2 μm), so we omitted the insulation layer between the conductive layers in the FEA modeling and used conductive blocks instead. The cross-section schematic diagram of the FEA model is shown in Figure S3. In the FEA of the bend cycling test, the displacement constraint was applied on either side of the plane, resulting in a bending angle of 120°. In the FEA of the thermal cycling test, simulation was conducted with the bending stress in order to fit the usage scenarios of conformal electronics. The thermal profile was consistent with the experiments, as shown in Figure 2c.

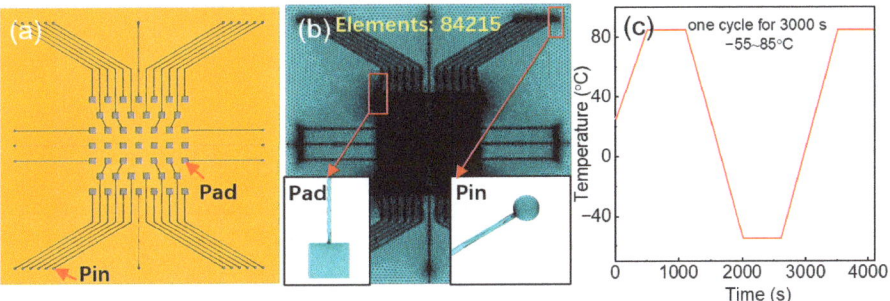

Figure 2. Pretreatment of finite element simulation: (**a**) FEA model; (**b**) mesh generation; (**c**) thermal cycling load files.

3. Results and Discussion

In high-precision printing, the line width is an important indicator for evaluating the printing effect. The width of the conductive path is also an important design parameter, as uncontrollable line widths would lead to short circuits between pins. Unlike traditional lithography processes, printing electronics cannot obtain specific line widths through masks, and the line widths acquired through printing have a strong process correlation. Therefore, it is necessary to clarify the rule of variation in line width through various process parameters to obtain conductive patterns with specific line widths. The influence of the printing speed, working height, and nozzle diameter of EHD printing on the line width is shown in Figure 3. As the printing speed increases, the line width gradually decreases, and the rate of change of the line width gradually decreases before stabilizing. The printing speed does not affect the diameter of the Taylor cone jet, but it does determine the amount of ink deposited per unit of time. A lower speed leads to greater ink deposition, and the wetting and spreading of the ink on the substrate increase the line width, as shown in Figure 3a. As the working height increases, the line width steadily increases due to the diffusion of the Taylor cone jet in the air, as shown in Figure 3b. When the working height approaches zero using high-speed printing, the line width becomes close to the diameter of the Taylor cone jet. The diameter of the nozzle directly determines the diameter of the Taylor cone jet. As the nozzle size increases, the line width steadily increases, as shown in Figure 3c. When the nozzle diameter is too large, it cannot produce a stable Taylor cone jet. Moreover, excessive printing speeds and working heights lead to discontinuous printing lines, resulting in large edge roughness of the printed lines. The diameter of the drill bit used for drilling in this study is 150 µm. The drill bit damages the conductive path when the line width is too small. We purposefully selected a parameter combination based on the designed line width of 200 µm, so we used a 300 µm nozzle, a working height of 100 µm, and a printing speed of 0.5 m/s in the subsequent multilayer circuit preparation process; using these parameters, we were able to reliably obtain a 200 µm line path, as shown in Figure S4 of the Supplementary Materials. The deposition rate of the ink is also a key parameter, and we calculated the ink deposition rate of 1.2 mL/h.

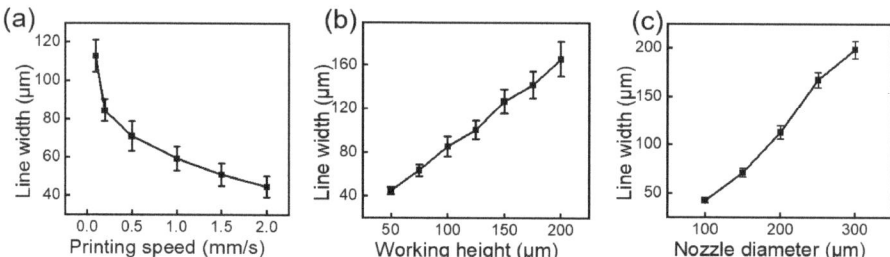

Figure 3. Printed line width under the influence of (**a**) printing speed, (**b**) working height, and (**c**) nozzle diameter.

Generally, metal nanoinks require post-treatment above 150 °C to exhibit good conductivity. However, the glass transition temperature of PET substrates is approximately 110 °C, so post-treatment needs to be carried out below 110 °C. In addition, we need to determine the conductivity mechanism, which is necessary to explain the increase in resistance during the bending cycles. The morphologies of the printed conductive path after 60 min of sintering at different temperatures are shown in Figure 4a–d; combined with the electrical resistivity results of Figure 4e, it can be seen that after post-treatment at temperatures below 150 °C, there was no significant change in the microstructure of the silver path, and micrometer sized silver nanoflakes were uniformly mixed as shown in Figure 4a–c. After sintering at 100 °C for 1 h, the resistance of the conductive circuit showed a significant decrease compared to the one without post-treatment, with a resistivity of 14.3 µΩ·cm. When the post-treatment temperature was raised to 150 °C, the resistance did not decrease further, indicating that the decrease in resistance caused the volatilization of organic solvents, which led to a decrease in the contact resistance between nanoflakes. However, at this time, these sliver nanoflakes were not sintered. When the temperature was raised to 300 °C, the resistance significantly decreased and quickly stabilized. The nanoflakes were sintered together due to the atomic diffusion at such a high temperature, which further reduced the resistivity, as shown in Figure 4d. In summary, sufficient conductivity could already be obtained at 100 °C via organic volatilization and the overlap between silver flakes, so we put the printed ten-layer circuit in a constant temperature box at 100 °C for 1 h.

Based on the process above, the ten-layer flexible circuits were prepared as shown in Figure 5. After testing the conductivity between different pins, effective interconnection was achieved between each layer. The unconnected pins were non-conductive, which proved that there was no short circuiting. The purpose of using a flexible substrate is to attach the circuit to a three-dimensional surface, so we attached the ten-layer circuits onto a wing-shaped resin substrate, as shown in Figure 5b,c. The resin block was fabricated using photocurable 3D printing, and the minimum curvature radius of the curved substrate was 20 mm. The flexible multilayered circuit showed an excellent conformal ability and tightly adhered to the curved substrate without delamination, proving that the multilayered circuit can adapt to complex curved surfaces.

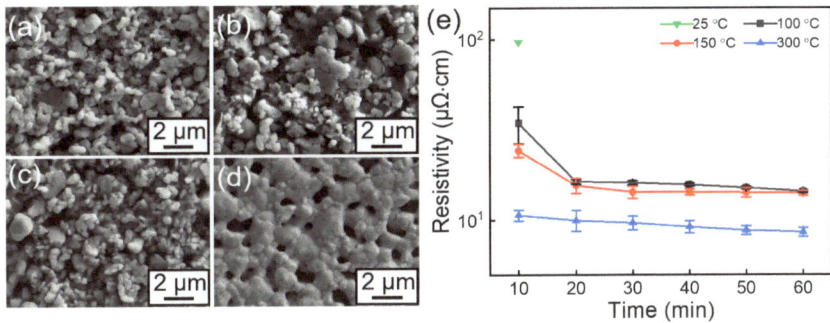

Figure 4. SEM images of the printed conductive path after 60 min sintering with the temperature of (**a**) 25 °C, (**b**) 100 °C, (**c**) 150 °C, and (**d**) 300 °C. (**e**) Variation curve of the electrical resistivity of conductive lines over time at different temperatures.

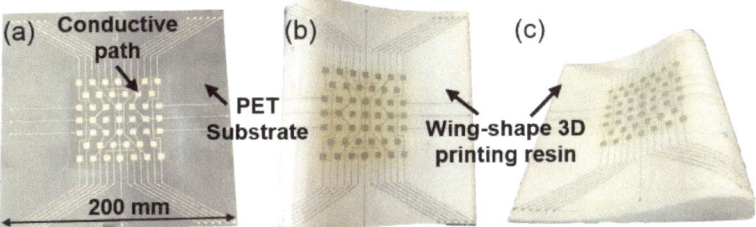

Figure 5. Optical images of (**a**) 10-layer flexible circuit and (**b**,**c**) conformal circuit.

Bending stress is the most common load on flexible circuits. To improve the bending reliability of the multilayered circuits, we established different conductive layer thicknesses in different areas in order to improve it. The part above the neutral plane bears the tensile stress during the bending process, while the part below the neutral plane bears the compressive stress. As mentioned in Figure 4, the good conductivity at a post-processing temperature of 100 °C is due to the overlap between the nanoflakes rather than a result of sintering. Compression stress will compress the nanoflakes to enhance conductivity. Therefore, the conductive layer can be printed thinner in the areas with higher compression stress. Conversely, tensile stress will loosen the nanoflakes, and an excessively thin conductive layer may result in an open circuit. Therefore, we made the conductive layer thicker in areas with higher tensile stress. We achieved a thicker conductive layer by increasing the number of prints, as shown in Figure 6a–c. We repeated printing twice for the area of the 10-layer circuit under compressive stress and obtained an average thickness of 5.6 μm. For the area under tensile stress, we repeated printing five times and obtained an average thickness of 26.5 μm, as shown in Figure 6d. A cross-section SEM image of a conductive through-hole is shown in Figure 6e. It can be seen that the conductive through-hole connected different conductive layers together. The interface between the conductive and insulating layers was tightly bonded without any delamination phenomenon, as shown in Figure 6f. EDS surface scanning was performed on the ten-layer circuit's cross-section, and the results are shown in Figure 6g–i. The main component of the conductive layer was Ag, and O and C elements were concentrated in the polyimide insulating layer, which separates the different conductive layers well and avoids the formation of short circuits. In addition, it can be found that there was no Ag in the insulating layer, but there was a low concentration of C and O elements in the conductive layer, which is due to the fact that we used a post-processing temperature of 100 °C and the organic compounds in the ink did not evaporate completely.

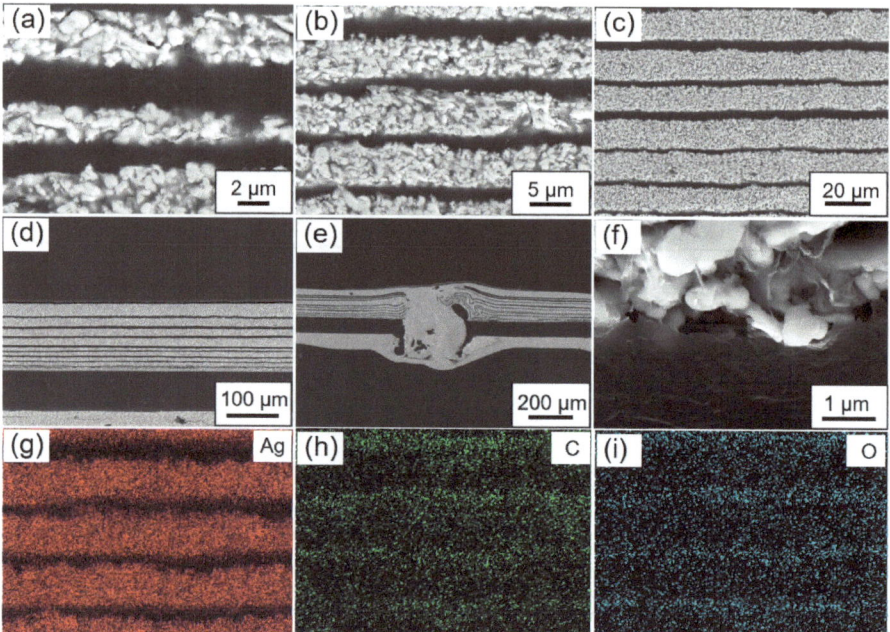

Figure 6. Cross-section SEM images of the different conductive layers thicknesses obtained via repeated printing (**a**) once, (**b**) twice, and (**c**) five times. Cross-section SEM images of (**d**) 10-layer flexible circuit, (**e**) conductive through-hole, and (**f**) interface between the conductive and insulating layers. EDS results of the 10-layer flexible circuit cross-section (**g**–**i**).

The curved substrates result in the flexible circuit undergoing bending stress for a long time. In order to verify the reliability of the flexible multilayer circuit under cyclic bending with a large curvature radius, a combination of simulations and experiments was carried out to explore the reliability and failure mechanism of the multilayer circuits. The displacement distribution and stress distribution are shown in Figure 7a,b, indicating that stress is concentrated in a small range of the bending center, only 0.55 MPa, due to the very small modulus of PET. A bend cycling test was conducted on flexible circuits to monitor the changes in resistance during the bending process, as shown in Figure 7c. The resistance exhibited periodic changes with increasing amplitude over the bending cycle. After 500 cycles, the resistance change rate was 7.9%, and after 8000 cycles, the resistance was approximately three times the initial resistance, demonstrating excellent bending reliability. Results of the microscopic morphology analysis of the circuit after 8000 bending cycles are shown in Figure 8. It can be seen that the microstructure on the surface of the circuit did not show obvious changes, but fine cracks appeared on the surface with a width of 1 μm. During the cyclic bending process, the maximum crack width was observed when the bending angle reached its maximum value and led to the maximum resistance. As the bending angle decreases, the nanoflakes around the cracks overlapped with each other, reducing the resistance and exhibiting periodic changes. Additionally, we found that, after the bend cycling experiment, the resistance of the circuit continued to slowly decrease. After 8000 bending cycles, the resistance was 2.6 times the initial resistance. However, the resistance decreased to 1.7 times the initial resistance after 10 h. This may be due to the effect of gravity, which makes the loose nanoflakes overlap in a tighter configuration.

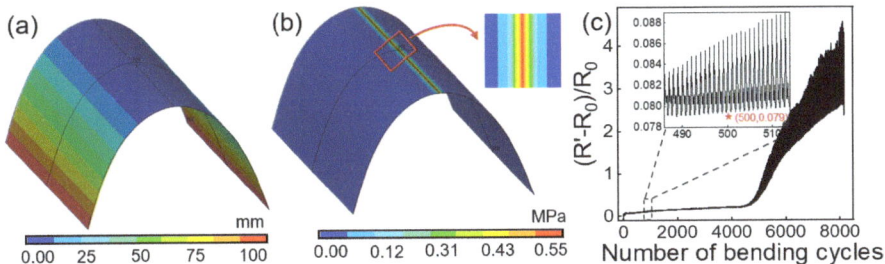

Figure 7. Bending reliability of the 10-layer flexible circuit: (**a**) displacement distribution obtained via FEA simulation; (**b**) von Mises stress distribution via FEA simulation; (**c**) resistance change rate during bending cycle test.

Figure 8. SEM morphology of the conductive pattern after 8000 bending cycles: (**a**) total area; (**b**) enlargement of the normal area; (**c**) enlargement of the crack area.

In order to make the multilayer circuit more compatible with a wider range of usage scenarios, the temperature cycling test environment was taken from −25 °C to 85 °C, and corresponding thermal cycling simulation was carried out. Conformal electronics experience both bending stress and thermal stress. Therefore, thermal–mechanical coupling simulations were carried out with bending prestress, and we applied the stress distribution obtained from bending simulation as a load, which can be used to explore the thermal cycling reliability of multilayer circuits under bending conditions. The temperature distribution at the moment of maximum temperature gradient during the heating process is shown in Figure 9a. The temperature distribution of the entire multilayer circuit was consistent, with a maximum temperature difference of only 0.13 °C. This is because the thickness of the multilayer circuit was only 0.2 mm, and the temperature quickly reached uniformity in a hot convection environment. The maximum stress during the thermal cycling process occurs at extremely low temperatures because the difference in the thermal expansion coefficient of the material is the greatest. Thermal stress is generated by mismatches in temperature, and the maximum stress point occurs at the junction of the silver line at the bending center and the PET, which was 6.8 MPa. It can be seen that the thermal stress generated by this multilayered circuit under thermal cycling load is extremely small. Due to the long testing time and the large amount of data required to detect changes in circuit resistance, multiple samples were tested simultaneously, and the initial resistance was measured as R_0. Samples were measure under 100, 200, 500, 800, and 1000 temperature cycles for resistance measurement, and the measured value was denoted as R'. The resistance change during thermal cycling is shown in Figure 9c. After 200 cycles, the resistance dropped to 95% of the initial resistance and remained unchanged. The initial slight decrease in resistance was because of the annealing effect in the high-temperature region, which further densified the conductive path. The experimental results were consistent with the simulation results, and the minimal thermal stress caused by thermal cycling did not cause a decrease in performance. Moreover, the resistance remained stable after 1000 thermal cycles.

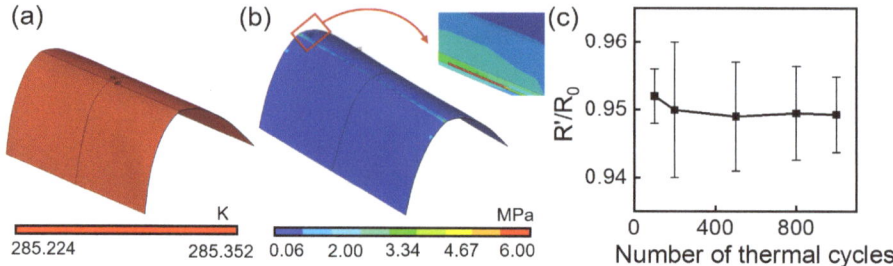

Figure 9. Thermal cycling reliability of the 10-layer flexible circuit. (**a**) Temperature distribution by FEA simulation. (**b**) Von Mises stress distribution by FEA simulation. (**c**) Resistance change rate during thermal cycling test.

4. Conclusions

This work proposed a large-area, low-cost, and highly reliable method for preparing multilayer flexible circuits. The advanced method of electrohydrodynamic printing was used to fabricate the conductive path, and the mechanisms by which various parameters influenced printing quality were explored in detail. Interlayer insulation was achieved by spraying PI layers and interconnection was achieved by micro vias. The resistivity of the silver conductive path after post-processing at 100 °C was only 14.3 µΩ·cm. The ten-layer flexible circuit has excellent bend cycling reliability and thermal cycling reliability. The failure mechanism under two cyclic loads was elucidated through a combination of numerical simulation and reliability testing. After 8000 bending cycles, the circuit resistance still met the high conductivity requirement, while the resistance only decreased by 5% after 1000 thermal cycles. The method proposed in this study has broad potential applications in the field of flexible devices, helping to accelerate the development of wearable electronics and conformal electronics.

Supplementary Materials: The following supporting information can be downloaded at: https://www.mdpi.com/article/10.3390/coatings14050625/s1.

Author Contributions: Methodology, S.W. (Shang Wang) and J.W.; validation, G.L. and J.F.; formal analysis, G.L.; investigation, S.W. (Shang Wang) and G.L.; writing—original draft preparation, G.L. and Y.S.; writing—review and editing, S.W. (Shang Wang) and S.W. (Shujun Wang); supervision, Y.T.; project administration, Y.T.; funding acquisition, Y.T. All authors have read and agreed to the published version of the manuscript.

Funding: This work was supported by the Heilongjiang Province Key Research and Development Program (Grant No. 2022XJ03C07), the Fundamental Research Funds for the Central Universities (Grant No. FRFCU5710051222), the Pre-Research Foundation of China (Grant No. 80923020702), and the State Key Laboratory of Precision Electronic Manufacturing Technology and Equipment.

Institutional Review Board Statement: Not applicable.

Informed Consent Statement: Not applicable.

Data Availability Statement: Data are contained within the article.

Conflicts of Interest: The authors declare no conflicts of interest.

References

1. Traglia, F.D.; Ciampalini, A.; Pezzo, G. Editorial: Synthetic aperture radar and natural hazards: Applications and outlooks. *Front. Earth Sci.* **2019**, *7*, 191. [CrossRef]
2. Wang, S.; Zeng, G.; Sun, Q.; Feng, Y.; Wang, X.; Ma, X.; Li, J.; Zhang, H.; Wen, J.; Feng, J.; et al. Flexible Electronic Systems via Electrohydrodynamic Jet Printing: A MnSe@rGO Cathode for Aqueous Zinc-Ion Batteries. *ACS Nano* **2023**, *17*, 13256–13268. [CrossRef] [PubMed]
3. Sun, Q.; Zeng, G.; Li, J.; Wang, S.; Botifoll, M.; Wang, H.; Li, D.; Ji, F.; Cheng, J.; Shao, H.; et al. Is Soft Carbon a More Suitable Match for SiO(x) in Li-Ion Battery Anodes? *Small* **2023**, *19*, e2302644. [CrossRef] [PubMed]

4. Jeon, Y.; Lee, D.; Yoo, H. Recent Advances in Metal-Oxide Thin-Film Transistors: Flexible/Stretchable Devices, Integrated Circuits, Biosensors, and Neuromorphic Applications. *Coatings* **2022**, *12*, 204. [CrossRef]
5. Ding, C.; Gao, X.; Tang, W.; Liu, H.; Zhu, C.; Bu, X.; An, J. A Balloon Conformal Array Antenna of Broadened Bandwidth, High Gain, and Steerable Directional Beam Fed by a Flexible Phase Shifting Network. *IEEE Trans. Antennas Propag.* **2023**, *71*, 8525–8536. [CrossRef]
6. Huang, Y.; Wu, H.; Zhu, C.; Xiong, W.; Chen, F.; Xiao, L.; Liu, J.; Wang, K.; Li, H.; Ye, D.; et al. Programmable robotized 'transfer-and-jet' printing for large, 3D curved electronics on complex surfaces. *Int. J. Extrem. Manuf.* **2021**, *3*, 045101. [CrossRef]
7. Shaw, M.; Choukiker, Y.K. Conformal microstrip array antenna with the omnidirectional pattern for 5G applications. *Microw. Opt. Technol. Lett.* **2022**, *64*, 2089–2094. [CrossRef]
8. Zhang, W.; Zhang, L.; Liao, Y.; Cheng, H. Conformal manufacturing of soft deformable sensors on the curved surface. *Int. J. Extrem. Manuf.* **2021**, *3*, 042001. [CrossRef]
9. Tan, H.W.; Choong, Y.Y.C.; Kuo, C.N.; Low, H.Y.; Chua, C.K. 3D printed electronics: Processes, materials and future trends. *Prog. Mater. Sci.* **2022**, *127*, 100945. [CrossRef]
10. Zavanelli, N.; Yeo, W.H. Advances in Screen Printing of Conductive Nanomaterials for Stretchable Electronics. *ACS Omega* **2021**, *6*, 9344–9351. [CrossRef] [PubMed]
11. Khan, Y.; Thielens, A.; Muin, S.; Ting, J.; Baumbauer, C.; Arias, A.C. A New Frontier of Printed Electronics: Flexible Hybrid Electronics. *Adv. Mater.* **2020**, *32*, e1905279. [CrossRef] [PubMed]
12. Li, Y.; Wang, R.; Zhu, X.; Yang, J.; Zhou, L.; Shang, S.; Sun, P.; Ge, W.; Xu, Q.; Lan, H. Multinozzle 3D Printing of Multilayer and Thin Flexible Electronics. *Adv. Eng. Mater.* **2022**, *25*, 2200785. [CrossRef]
13. Sun, P.; Zhang, S.; Zhu, X.; Li, H.; Li, Y.; Yang, J.; Peng, Z.; Zhang, G.; Wang, F.; Lan, H. Directly Printed Interconnection Wires between Layers for 3D Integrated Stretchable Electronics. *Adv. Mater. Technol.* **2022**, *7*, 2200302. [CrossRef]
14. Cai, S.; Sun, Y.; Wang, Z.; Yang, W.; Li, X.; Yu, H. Mechanisms, influencing factors, and applications of electrohydrodynamic jet printing. *Nanotechnol. Rev.* **2021**, *10*, 1046–1078. [CrossRef]
15. Chen, J.; Wu, T.; Zhang, L.; Song, H.; Tang, C.; Yan, X. Flexible conductive patterns using electrohydrodynamic jet printing method based on high-voltage electrostatic focusing lens. *Int. J. Adv. Manuf. Technol.* **2023**, *127*, 4321–4329. [CrossRef]
16. Ma, J.; Feng, J.; Zhang, H.; Hu, X.; Wen, J.; Wang, S.; Tian, Y. Electrohydrodynamic Printing of Ultrafine and Highly Conductive Ag Electrodes for Various Flexible Electronics. *Adv. Mater. Technol.* **2023**, *8*, 2300080. [CrossRef]
17. Bi, S.; Wang, R.; Han, X.; Wang, Y.; Tan, D.; Shi, B.; Jiang, C.; He, Z.; Asare-Yeboah, K. Recent Progress in Electrohydrodynamic Jet Printing for Printed Electronics: From 0D to 3D Materials. *Coatings* **2023**, *13*, 1150. [CrossRef]
18. Xu, R.; He, P.; Lan, G.; Behrouzi, K.; Peng, Y.; Wang, D.; Jiang, T.; Lee, A.; Long, Y.; Lin, L. Facile Fabrication of Multilayer Stretchable Electronics via a Two-mode Mechanical Cutting Process. *ACS Nano* **2022**, *16*, 1533–1546. [CrossRef] [PubMed]
19. Wang, Y.; Xu, C.; Yu, X.; Zhang, H.; Han, M. Multilayer flexible electronics: Manufacturing approaches and applications. *Mater. Today Phys.* **2022**, *23*, 100647. [CrossRef]
20. Phung, T.H.; Jeong, J.; Gafurov, A.N.; Kim, I.; Kim, S.Y.; Chung, H.-J.; Kim, Y.; Kim, H.-J.; Kim, K.M.; Lee, T.-M. Hybrid fabrication of LED matrix display on multilayer flexible printed circuit board. *Flex. Print. Electron.* **2021**, *6*, 024001. [CrossRef]
21. Song, S.; Hong, H.; Kim, K.Y.; Kim, K.K.; Kim, J.; Won, D.; Yun, S.; Choi, J.; Ryu, Y.I.; Lee, K.; et al. Photothermal Lithography for Realizing a Stretchable Multilayer Electronic Circuit Using a Laser. *ACS Nano* **2023**, *17*, 21443–21454. [CrossRef] [PubMed]
22. Jeong, S.; Kim, T.-W.; Lee, S.; Sim, B.; Park, H.; Son, K.; Son, K.; Kim, S.; Shin, T.; Kim, Y.-C.; et al. Analysis of Repetitive Bending on Flexible Wireless Power Transfer (WPT) PCB Coils for Flexible Wearable Devices. *IEEE Trans. Compon. Packag. Manuf. Technol.* **2022**, *12*, 1748–1756. [CrossRef]
23. Seo, Y.; Ha, H.; Matteini, P.; Hwang, B. A Review on the Deformation Behavior of Silver Nanowire Networks under Many Bending Cycles. *Appl. Sci.* **2021**, *11*, 4515. [CrossRef]
24. Li, B.; Zhou, W.; Hu, Y.; Du, M.; Wang, M.; Zhang, D.; Wang, H. Statistical Study of Degradation of Flexible Poly-Si TFTs Under Dynamic Bending Stress. *IEEE J. Electron Devices Soc.* **2022**, *10*, 123–128. [CrossRef]

Disclaimer/Publisher's Note: The statements, opinions and data contained in all publications are solely those of the individual author(s) and contributor(s) and not of MDPI and/or the editor(s). MDPI and/or the editor(s) disclaim responsibility for any injury to people or property resulting from any ideas, methods, instructions or products referred to in the content.

Article

Study on the Solder Joint Reliability of New Diamond Chip Resistors for Power Devices

Wenyu Wu [1], Geng Li [2], Shang Wang [2,*], Yiping Wang [2], Jiayun Feng [2], Xiaowei Sun [1] and Yanhong Tian [2,*]

[1] NO. 38 Research Institute, China Electronics Technology Group Corporation, Hefei 230031, China; hzhcandy@126.com (W.W.)
[2] State Key Laboratory of Advanced Welding and Joining, Harbin Institute of Technology, Harbin 150001, China; li-geng@hit.edu.cn (G.L.)
* Correspondence: wangshang@hit.edu.cn (S.W.); tianyh@hit.edu.cn (Y.T.)

Abstract: New diamond chip resistors have been used in high-power devices widely due to excellent heat dissipation and high-frequency performance. However, systematic research about their solder joint reliability is rare. In this paper, a related study was conducted by combining methods between numerical analysis and laboratory reliability tests. In detail, the shape simulation and thermal cycling finite element simulation for solder joints with different volumes were carried out. The optimized solder volume was 0.05 mm^3, and the maximum thermal cycling stress under the optimized shape was 38.9 MPa. In addition, the thermal cycling tests with current and high temperature storage tests were carried out for the optimized solder joint, which showed good agreement with the simulation results, clarified the growth and evolution law of intermetallic compound at the interconnection interface, and proved the optimized solder joint had great anti-electromigration, temperature cycling and high temperature storage reliability. In this work, an optimized solder joint structure of a diamond chip resistor with high reliability was finally obtained, as well as providing considerable reliability data for the new type of diamond chip resistors, which would boost the development of power devices.

Keywords: diamond chip resistor; solder joint; reliability; numerical analysis

1. Introduction

High-power electronic devices are widely used in the field of wireless communication [1,2]. With the demand for high performance and light weight of electronic devices, the continuous improvement of their integration has brought a significant increase in power density and the large current and high heat have led to many reliability problems [3]. Numerous studies have shown that most of the failures of electronic devices come from the interconnection solder joints of electronic packaging due to the stress caused by the thermal mismatch of materials and the evolution of microstructure and morphology [4]. As the most common component of electronic devices, chip resistors are used in almost every electronic device. Therefore, ensuring the reliability of chip resistor solder joints under high temperature and power cycling is the cornerstone of the development of high-power devices.

With continuous power and frequency improvement of power devices, the commonly used materials Al_2O_3 of chip resistor substrate cannot meet the requirements of heat dissipation [5,6]. The thermal conductivity of CVD diamond is 2000 W·m^{-1}·K^{-1}, which is dozens of times higher than Al_2O_3. Using diamond as the substrate material of chip resistors can greatly improve the heat dissipation performance [6]. In addition, this new type of diamond chip resistor has advantages of high frequency (more than 30 GHz), high power, small size, light weight and stable performance, and can service in extreme environments such as deep space exploration and military equipment that require ultra-high reliability [7]. The usage of new materials always leads to new reliability problems. There have been

many studies on the reliability of traditional chip resistor solder joints [8,9], but the research on the reliability of new chip resistor solder joints is not sufficient. Therefore, it is of great significance to supplement the reliability data of the new diamond chip resistors to ensure the stable operation of power devices and assist the development of power devices.

Due to the frequent change of ambient temperature or switch of power devices, chip resistors need to undergo temperature cycling. Under the alternating temperature load, the thermal mismatches of different interfaces of materials generate large thermal stress, which leads to plastic deformation, grain boundary slip and grain boundary defects in solder joints. Stress concentrates at those defects and results in initial cracks, then cracks propagate to cause failure [10]. It is necessary to study the reliability of new chip resistors under temperature cycling. The finite element analysis method can save a lot of time and cost in the reliability analysis process, which can quickly obtain the stress and strain response of the device under a certain load and predict the service life of solder joints by combining a constitutive equation and life prediction model [11–15]. The accuracy of the finite element model greatly affects the accuracy of simulation and the simulation of the solder joint shape plays an important role in improving the accuracy of model. In addition, the shape of the solder joint directly determines the stress distribution of the solder joint. Therefore, the solder joint shape with the lowest stress level can be obtained through the iterative optimization of solder joint shape simulation and thermal cycle finite element simulation.

The new diamond chip resistors in power devices need to work at a high temperature due to high power. The continuous high temperature causes excessive growth of brittle intermetallic compounds (IMCs) at the soldering interface, which leads to an increasing risk of failure. The mechanism of IMCs growth involves the diffusion and migration of elements. The diffusion of elements is affected by many factors such as element concentration, temperature, stress, etc. [16], therefore, it is difficult to evaluate the reliability of solder joints only by simulation, so the high temperature storage (HTS) experiment is essential.

In order to realize the electrical connection of the diamond chip resistors, it is necessary to electroplate electrodes at the terminals of the resistor. The electrode material is usually nickel. The quality of the nickel coating will significantly affect the reliability of the solder joint in the process of temperature cycling or aging [17]. Therefore, this paper also focuses on the analysis of the reliability of the coating interface.

With the background mentioned above, this study conducted the solder joints shape simulation and thermal cycling finite element simulation for the new diamond chip resistors. In order to verify the simulation results, corresponding thermal cycling tests were carried out. The power devices also bore current load during service, so another control group with current was set in order to evaluate the impact of current on solder joints' reliability. In addition, high temperature storage tests were carried out to evaluate the reliability of the solder joints of the new diamond chip resistors stored under high temperature for a long time.

2. Materials and Methods

2.1. Numerical Simulation

In order to investigate the device-level reliability, the power divider with new diamond chip resistors was selected as the study object. Several assumptions were made to ensure the accuracy and feasibility of the numerical simulation as listed below:

- The metal plating was ignored;
- All materials were uniform and dense;
- All interconnecting interfaces were tightly combined;
- The effect of gravity was considered in the simulation of solder joint shape;
- The change of material thermodynamic parameters with temperature was considered in the thermal cycling simulation.

The process of numerical simulation was shown in Figure 1. Parameterized modeling was carried out in the Surface Evolver software (version 2.70) according to the size data of the chip resistor. The differential equation of surface wetting was set to calculate the solder joint shape under different solder volumes by the minimum energy principle. Then, the Surface Evolver (SE) model of the chip resistor was imported into ANSYS (version 14.0), and the parametric modeling of the whole power divider was carried out. The Anand model was used to describe the viscoplastic behavior of solder under a low stress state. The thermodynamic simulation used thermal element SOLID70 and structural element SOLID185. Tetrahedral elements were used for meshing due to plenty of irregular structures, and the mesh of the solder part was refined to improve the simulation accuracy. Fixed constraints were applied at the bolt holes at the four corners of the power divider shell, as shown in Figure 2a. The thermal cycling profile was in accordance with the requirements of TC4 in IPC9701. The temperature was from −55 to 125 °C, one cycle for 50 min, whose rising time and falling time were 15 min. Some research showed the inelastic strain amplitude of the solder joint was generally stable after 7 to 8 cycles [11–13], so a total of 10 thermal cycles were applied to the model, as shown in Figure 2b. The material parameters used in the numerical simulation can be found in Tables S1 and S2 of the Supplementary Materials.

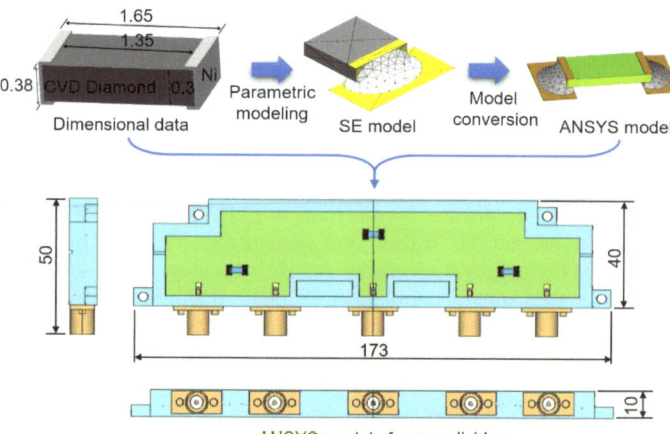

Figure 1. Schematic process of numerical simulation.

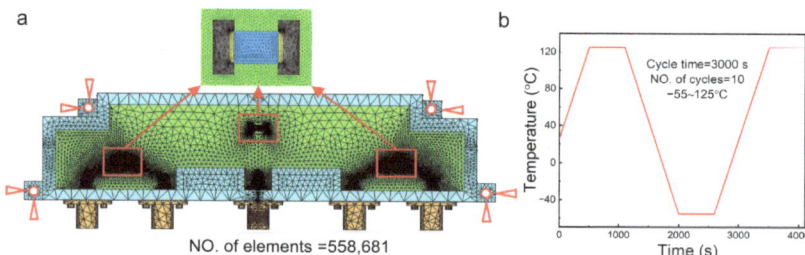

Figure 2. Pretreatment of finite element simulation: (**a**) Mesh generation and constraints; (**b**) Load files.

2.2. Laboratory Experiment

Screen printing and reflow soldering process were carried out according to the optimal solder volume obtained by simulation, and the new diamond chip resistors were soldered onto the microstrip plate of power divider. The chip resistors were the CRD0603DX5W2 model produced by Smiths Interconnect (London, UK) and the solder was eutectic SnPb. The corresponding thermal cycling tests were carried out on the power divider by high-low temperature test-box to verify the simulation results, and the samples after 200, 500, and 1000 cycles were taken out for analysis. In order to determine the anti-electric migration ability of the solder joint, another energized control group was set with the current of 0.15 A. The aging temperature of HTS test was set to 150 °C by constant temperature drying box, and the temperature accuracy was ±0.1 °C. The growth rate of intermetallic compounds followed Fick's diffusion law and was proportional to the quadratic power of time, so aging times of the samples were set to be 1 day, 4 days, 9 days, 16 days, 25 days, 36 days, and 49 days.

The obvious hardness difference made it difficult to obtain the flat interconnection interface between the diamond and SnPb solder, so the interface was polished with a cross-section polisher. The direction of the ion beam was perpendicular to the plane of the microstrip plate and polished 6 h for each sample. The microstructure of the solder joints was investigated by the scanning electron microscopy (SEM, ZEISS Sigma 300, Jena, Germany), and the SEM images were taken by backscattered electrons (BE) to obtain compositional contrast. The composition of IMCs were identified by energy dispersive spectrometer (EDS, ZEISS Sigma 300), and the thickness measurement of interface IMCs was calculated by the Image J (version 1.8.0) according to the imaging contrast difference of different materials and the thickness was an average value of all samples under same conditions. We took three chip resistors for each test condition to avoid contingency.

3. Results and Discussion

The simulation results of solder joint shape are shown in Figure 3, where V is the volume of solder. The solder was only at the bottom of the resistor with the volume of 0.01 mm^3, and the lateral wall of the resistor was not wet. As the volume of solder increased, the solder began to wet the lateral wall of the resistor and continued to climb. When the volume came to 0.025 mm^3, solder climbed more than half the height of the resistor. The solder climbed to the top of the resistor when the volume became 0.05 mm^3, but the width direction of the resistor was not completely wetted and the solder joint presented a fillet shape of chip components as usual at this time. When the volume of the solder increased further, the solder continued to wet on the pad and the direction of resistor width until the volume of solder reached 0.13 mm^3, the interfacial tension between solder and pad could not resist the gravity of the solder at this time, so the solder collapsed and overflowed the pad area. According to the SMT general test standard IPC-610F, the solder of chip components should be higher than 1/2 of the component's height, so we took the solder joints with the volume of 0.025 mm^3, 0.05 mm^3, 0.075 mm^3, 0.01 mm^3, and 0.025 mm^3 for subsequent thermal cycling simulation.

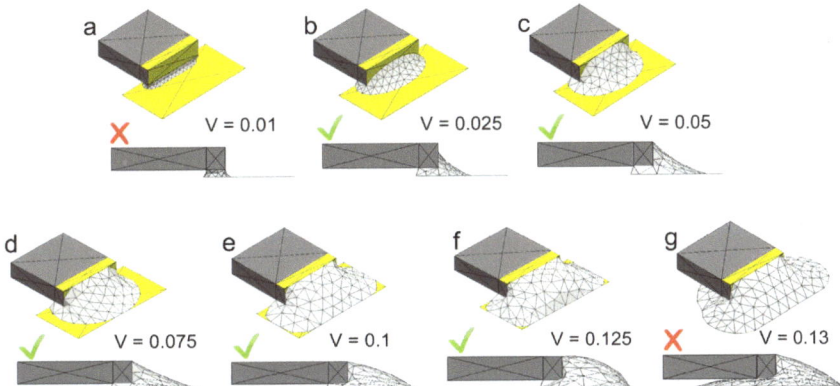

Figure 3. Simulation results of solder joint shape under the different solder volumes of (**a**) 0.01 mm^3, (**b**) 0.025 mm^3, (**c**) 0.05 mm^3, (**d**) 0.075 mm^3, (**e**) 0.1 mm^3, (**f**) 0.125 mm^3, (**g**) 0.13 mm^3.

The thermal cycling simulation was taken for solder joints with different solder volumes in power divider with chip resistors. Among the three chip resistors carried by the power divider, the thermal stress of the solder joints near four-corner screw holes was the highest because they were close to the fixed constraint and had difficulty releasing the stress through deformation. During a whole thermal cycle, stress of the joints increased to the highest when the temperature just fell to minimum, because the CTE of each material had a larger difference in the low temperature. The stress distribution of the solder joint near the screw hole at the lowest temperature is shown in Figure 4. The maximum von Mises stress and strain were produced at the corner where the inner side of the solder joint contacted with the lower edge of the resistor. Because the horizontal distance between the inner side of solder pad and the resistor was very small, the solder joint there showed a nearly right angle and the stress was concentrated at this part. The maximum stress of solder joints of different shapes was between 39.0 and 43.4 MPa, and there was no significant difference. Although the maximum stress was slightly higher than the tensile strength of SnPb solder (36 MPa), the high stress here was only concentrated in a very short time and a very small area, after which the stress could be released through microcracks here, which did not affect the overall reliability of solder joints.

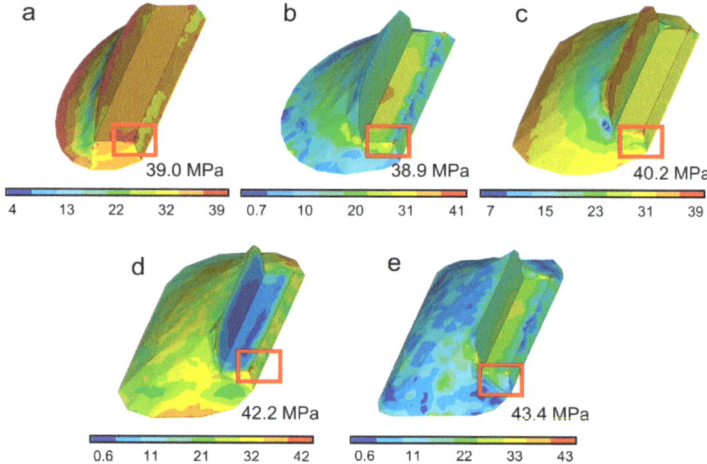

Figure 4. Stress distribution of solder joints at the lowest temperature with the solder volume of (**a**) 0.025 mm^3, (**b**) 0.05 mm^3, (**c**) 0.075 mm^3, (**d**) 0.1 mm^3, (**e**) 0.125 mm^3.

In Figure 4a, the solder joint was too small to release the stress through deformation of the solder joint, resulting in the highest-level overall stress when the solder volume was 0.025 mm^3. The stress level of solder joints with solder volume of 0.075 and 0.01 mm^3 was also large, which was due to the terrible shape of solder joints. The overall stress of solder joints with solder volume of 0.05 and 0.0125 mm^3 was much lower, but too much solder may lead to a Manhattan effect of the chip resistor during reflow soldering. Through the above analysis, the solder joint with a volume of 0.05 mm^3 was considered to be the best.

The stress-strain curve at the critical point of the solder joint with a volume of 0.05 mm^3 is shown in Figure 5. The curve was in the shape of a gradually convergent hysteresis loop, and the area surrounded by it represented the plastic strain energy. In further thermal cycling, the stress of the critical point was basically consistent and the plastic strain was accumulating at the same temperature in each cycle. After five cycles, the cumulative plastic strain of each cycle reached stability, and the value of the inelastic strain amplitude in one cycle after stabilization was 0.00456.

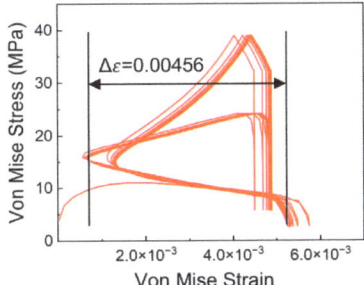

Figure 5. Stress-strain relationship of the critical point during ten thermal cycles.

Among many solder joint life prediction models, the Coffin–Manson equation with correction terms has a higher accuracy of low cycle fatigue life prediction, because it takes the conditions of thermal cycling into account such as the temperature amplitude and frequency. The modified Coffin–Manson model is the most widely-used model in the field of electronic packaging at present. The general expression of this model is [14,15]:

$$N_f = \frac{1}{2}(\frac{\Delta \gamma}{2\varepsilon_f})^{\frac{1}{c}}, \quad (1)$$

where N_f is the number of cycles to fail; $\Delta \gamma$ is the range of equivalent inelastic shear strain; ε_f is the fatigue ductility coefficient which is approximately equal to 0.325 for SnPb solder; c is the fatigue ductility index, a constant related to the thermal cycle temperature, equal to 3.39 for this work. The range of equivalent inelastic strain can be extracted from the finite element simulation results, which has the following relationship with $\Delta \gamma$:

$$\Delta \gamma = \sqrt{3} \Delta \varepsilon, \quad (2)$$

where $\Delta \varepsilon$ is the range of equivalent inelastic strain.

The thermal cycling life of solder joint was calculated by taking $\Delta \varepsilon$ into the formulas above, and the calculated life was 5.2×10^4 cycles, which was not the typical low cycle fatigue failure. It could be considered that this optimized solder joint structure would not have fatigue failure cause by creep deformation during thermal cycling, and has a great thermal cycling reliability.

The cross-section of the solder joints after the thermal cycling test are shown in Figure 6. Within 1000 cycles, the solder joints still had no microcracks, voids, or other defects, which verified the simulation results and confirmed that the optimized structure of the new diamond chip resistor solder joint had good thermal cycling reliability. With the increase in temperature cycles, the eutectic phase of solder was only slightly coarsened, which would not significantly affect the mechanical properties of solder joints.

The Pb element in eutectic SnPb does not participate in the interface reaction. According to the Cu-Sn binary phase diagram, the IMCs can only be Cu_6Sn_5 and Cu_3Sn below 350 °C. In addition, numerous studies have shown that during thermal cycling and aging, the reaction products of Cu-Sn interface are Cu_3Sn near the side of Cu and Cu_6Sn_5 near the side of Sn [18–22]. Considering the significant difference in atomic ratios between the two compounds and the accuracy of EDS, we used EDS to determine the type of IMCs. The EDS results can be found in Figure S1 of the Supplementary Materials. The IMC of solder–pad interface near the Cu pad was continuous Cu_3Sn where the Cu-Sn atomic ratio of EDS was very close to 3:1, while near the SnPb was continuous Cu_6Sn_5 with a microstructure of scallop where the atomic ratio was very close to 6:5. The IMC of solder–resistor interface was continuous layered $(Cu, Ni)_6Sn_5$ due to similar crystal structure of Cu and Ni [19].

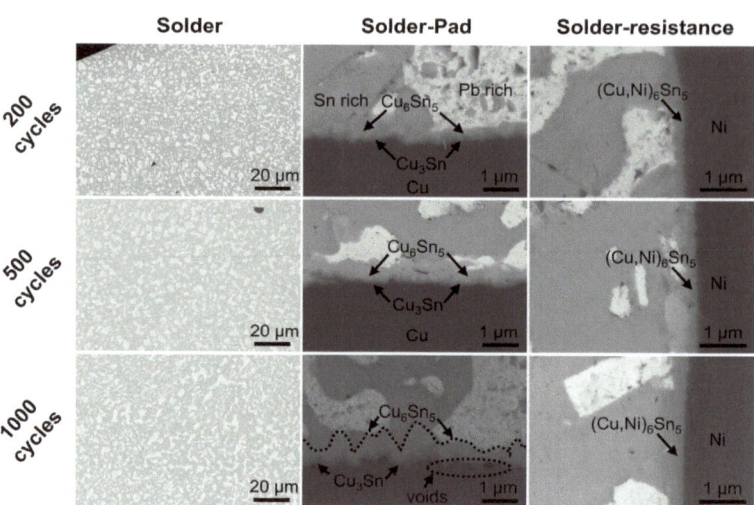

Figure 6. Cross-section SEM images of the solder joints after different thermal cycles.

As the thermal cycling test continued, the thickness of Cu_6Sn_5 and Cu_3Sn at the solder–pad interface increased obviously. It was because high temperature promoted the diffusion of elements, and Cu atoms reacted with Sn atoms to generate Cu_6Sn_5, and Cu_6Sn_5 was converted into Cu_3Sn gradually [18,19]. Kirkendall voids with diameter of 100–400 nm formed in the Cu_3Sn layer and along the Cu_3Sn/Cu interface. Those voids were generated because the diffusion rate of Cu in Cu_3Sn was much higher than that of Sn, resulting in the accumulation of excess vacancies [20,21]. The interface strain caused by lattice mismatch or the elastic anisotropy between phases was another factor that led to the interface voids [22]. The thickness of IMC in the solder–resistor interface $(Cu, Ni)_6Sn_5$ is significantly lower than that of the solder–pad interface, which is due to the slow reaction of the main reaction elements Ni and Sn at the solder–resistor interface.

Current will cause the directional migration of atoms, which accelerates or inhibits the growth of IMCs, resulting in microcracks, voids, and other defects [23,24]. The mechanism of the influence of the current on the reliability is shown in Figure 7. A is the IMC of the solder–pad interface at the side where electrons flow out, B is the IMC of the solder–pad interface at the side where electrons flow in, C is the IMC of the resistor–solder interface at the side where electrons flow out, and D is the IMC of the resistor–solder interface at the side where electrons flow in. Because electrons are conducive to the migration of Cu atoms at B and Ni atoms at C, the IMCs at B and C are generally thicker than that at A and D.

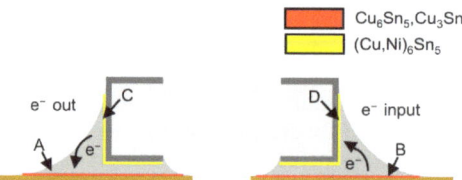

Figure 7. Diagram of the influence of current.

The section of the solder joints of the electrified samples were analyzed and there were no microcracks, voids, or other defects. The microstructure of the solder joints was basically the same as that of the untreated solder joints. In order to evaluate the effect of the current on the IMCs growth, the IMCs thickness under the different electrified conditions were calculated as shown in Table 1. The thickness of IMCs at A and D were basically the same no matter whether the solder joint was electrified or not. Although the electron flow was not conducive to the migration of the main diffusion atoms here, the thermal effect generated by the current promotes the diffusion to a certain extent. The IMCs at B and C under electrified condition were thicker than that of the untreated solder joints. Within 1000 cycles, the difference of IMC thickness at B between the electrified and untreated solder joints was within 0.1 µm, while D was within 0.2 µm. It was proved that the solder joints' structure had excellent reliability against electromigration.

Table 1. The thickness of IMCs at different interfaces under thermal cycling.

NO. of Cycles	Current	Thickness of IMCs (µm)			
		A	B	C	D
200	off	0.31	-	0.24	-
500	off	0.49	-	0.27	-
1000	off	0.72	-	0.32	-
200	on	0.30	0.34	0.35	0.27
500	on	0.51	0.59	0.44	0.31
1000	on	0.73	0.80	0.49	0.38

The sections of solder–pad interface with different aging time were shown in Figure 8. It can be seen that there were no cracks and delamination at the interconnection interface within 49 days of aging. The IMC of the unaged sample solder–pad interface was scallop-shaped Cu_6Sn_5, which was generated by the liquid phase reaction between the molten SnPb solder and the pad during reflow soldering. In the subsequent aging process, the reaction $Cu_6Sn_5 + 9Cu = 5Cu_3Sn$ occurred at the Cu_6Sn_5-Cu interface. The IMC thickness increased with the aging time. That was because the high temperature promoted atomic diffusion, and the growth of IMC was then dominated by atomic diffusion [25–29]. Cu_6Sn_5 grains ripen with the prolongation of aging time, large grains swallowed up small grains, grains became coarse, the number of grains decreased, and the IMC gradually became layered type [30–34].

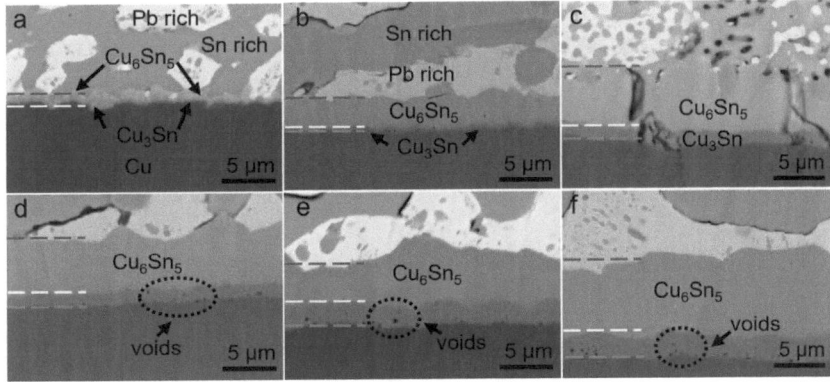

Figure 8. Cross-section SEM images of the solder–pad interface under the aging of (**a**) 0 day, (**b**) 4 days, (**c**) 16 days, (**d**) 25 days, (**e**) 36 days, (**f**) 49 days.

The relationship between IMCs thickness and aging time at the solder–pad interface is shown in Figure 9. The thickness of IMCs was proportional to the square root of aging time. The thickness of Cu_6Sn_5 and Cu_3Sn at 150 °C met the following formulas:

$$h_{Cu_6Sn_5} = 0.994t^{0.5} + 0.453, \qquad (3)$$

$$h_{Cu_3Sn} = 0.4t^{0.5} + 0.1, \qquad (4)$$

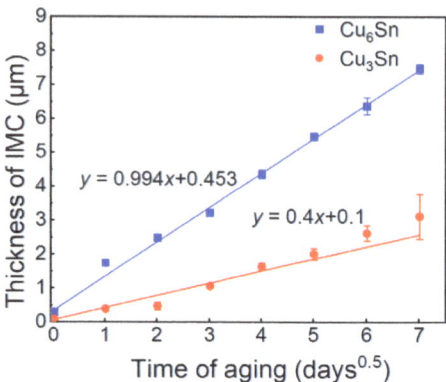

Figure 9. Curve fitting of IMC growth at solder–pad interface.

Through curve fitting, where h is the thickness of IMC, t is the time of aging. It was verified that IMC growth was dominated by element diffusion during aging and followed Fick's diffusion law. The IMC thickness at any temperature and time could be calculated by this fitting formula combining with the diffusion activation energy and the Arrhenius formula to evaluate the reliability of the solder joints at HTS.

Kirkendall voids appeared at the Cu_3Sn–Cu interface during aging. The atomic diffusion rate was higher than the temperature cycling, so Kirkendall voids were much more, but they did not gather to form cracks, which did not affect the reliability. Within 49 days of aging, there were no cracks, interface delamination, or other defects, and the eutectic phase of the solder was slightly coarsened with the aging time. The IMC at the solder–resistor interface was continuous layered $(Cu, Ni)_6Sn_5$, shown as Figure 10. At the beginning of aging process, the thickness of $(Cu, Ni)_6Sn_5$ increased, the Ni layer at the side of chip resistor was continuously consumed, and the Ni layer had completely reacted after 16 days of aging. At this time, the thickness of $(Cu, Ni)_6Sn_5$ did not increase significantly with the aging time, and the interface bonding was always tight, maintaining good reliability.

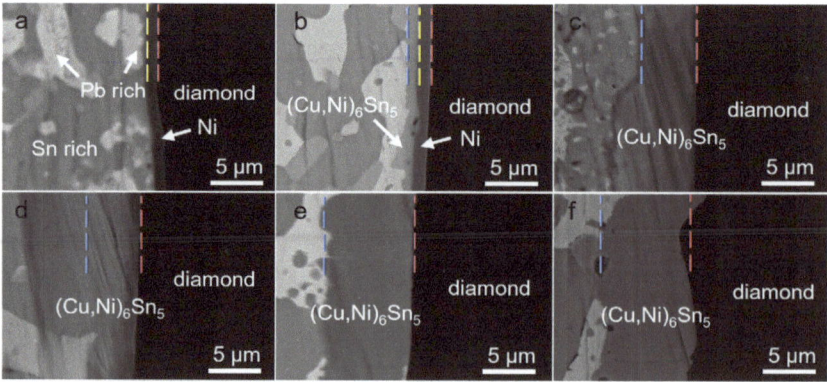

Figure 10. Cross-section SEM images of the solder–resistor interface under the aging of (**a**) 0 day, (**b**) 4 days, (**c**) 16 days, (**d**) 25 days, (**e**) 36 days, (**f**) 49 days.

4. Conclusions

This work systematically studied the reliability of the new chip diamond resistor solder joints, proposed a method to optimize the shape of the solder joints through numerical simulation, and determined the optimal solder volume of 0.05 mm^3, which was verified to be accurate by experiments. This optimized solder joint could guide the welding process of new chip resistors and improve the thermal cycling reliability of devices. The micro-morphology change, IMCs growth, and evolution law of the new chip resistor solder joints under different loads were determined through a series of reliability tests. During the temperature cycling, the growth of IMCs was relatively slow, while under 2A current, the IMC growth was slightly promoted. The IMC thickness difference between the two sides of the chip resistor solder joint was within 0.2 μm, and IMCs grew faster during the aging process. The growth pattern of Cu_6Sn_5 and Cu_3Sn were obtained by curve fitting, and the growth rate of IMCs at different temperatures could be predicted combined with the Arrhenius equation to evaluate the reliability. The results showed that the optimized solder joints had excellent thermal cycling reliability, anti-electric migration reliability, and HTS reliability, which could not only guide the welding process of new diamond chip resistors, but also accumulate some reliability data for it.

Supplementary Materials: The following supporting information can be downloaded at: https://www.mdpi.com/article/10.3390/coatings13040748/s1, Figure S1: EDS results of IMCs; Table S1: material properties at 25 °C; Table S2: Anand's constants of SnPb solder.

Author Contributions: Methodology, W.W. and X.S.; validation, G.L. and J.F.; formal analysis, G.L.; investigation, W.W. and G.L.; writing—original draft preparation, G.L.; writing—review and editing, S.W. and Y.W.; supervision, Y.T.; project administration, W.W. and Y.T.; funding acquisition, Y.T. All authors have read and agreed to the published version of the manuscript.

Funding: This research was supported by Heilongjiang Touyan Innovation Team Program (Grant No. HITTY-20190013) and National Natural Science Foundation of China (Grant No. U2241223).

Institutional Review Board Statement: Not applicable.

Informed Consent Statement: Not applicable.

Data Availability Statement: The datasets used and/or analysis results obtained in the current study are available from the corresponding author upon request.

Conflicts of Interest: The authors declare no conflict of interest.

References

1. Traglia, F.D.; Ciampalini, A.; Pezzo, G. Editorial: Synthetic aperture radar and natural hazards: Applications and outlooks. *Front. Earth Sci.* **2019**, *7*, 1–2. [CrossRef]
2. Dinc, T.; Zihir, S.; Gurbuz, Y. CMOS SPDT T/R switch for X-band, on-chip radar applications. *Electron. Lett.* **2010**, *46*, 1382–1384. [CrossRef]
3. Yao, Y.; Long, X.; Keer, L.M. A review of recent research on the mechanical behavior of lead-free solders. *J. Phys. D Appl. Mech. Rev.* **2017**, *69*, 040802. [CrossRef]
4. Jiang, N.; Zhang, L.; Liu, Z.; Sun, L.; Long, W.; He, P.; Xiong, M.; Zhao, M. Reliability issues of lead-free solder joints in electronic devices. *Sci. Technol. Adv. Mat.* **2020**, *20*, 876–901. [CrossRef]
5. Donato, N.; Rouger, N.; Pernot, J.; Longobardi, G.; Udrea, F. Diamond power devices: State of the art, modelling, figures of merit and future perspective. *J. Phys. D Appl. Phys.* **2020**, *53*, 093001. [CrossRef]
6. Yamasaki, S.; Makino, T.; Takeuchi, D.; Ogura, M.; Kato, H.; Matsumoto, T.; Iwasaki, T.; Hatano, M.; Suzuki, M.; Koizumi, S.; et al. Potential of diamond power devices. In Proceedings of the 2013 25th International Symposium on Power Semiconductor Devices and ICs, Kanazawa, Japan, 26 May 2013.
7. Bailly, M. Diamond RF (TM) resistives: The answer to high power and low capacitance. *Microw. J.* **2010**, *53*, 94.
8. Han, C.; Han, B. Board level reliability analysis of chip resistor assemblies under thermal cycling: A comparison study between SnPb and SnAgCu. *J. Mech. Sci. Technol.* **2014**, *28*, 879–886. [CrossRef]
9. Seo, W.; Ko, Y.; Yoo, S. Void fraction of a Sn–Ag–Cu solder joint underneath a chip resistor and its effect on joint strength and thermomechanical reliability. *J. Mater. Sci-Mater. El.* **2019**, *30*, 15889–15896. [CrossRef]
10. Che, F.X.; Pang, J.L. Fatigue reliability analysis of Sn-Ag-Cu solder joints subject to thermal cycling. *IEEE T. Device Mat. Re.* **2013**, *13*, 36–49. [CrossRef]
11. Che, F.X.; Pang, J.L. Vibration reliability test and finite element analysis for flip chip solder joints. *Microelectron. Reliab.* **2009**, *49*, 754–760. [CrossRef]
12. Chen, Z.; Zhang, Z.; Dong, F.; Liu, S.; Liu, L. A Hybrid finite element modeling: Artificial neural network approach for predicting solder joint fatigue life in wafer-level chip scale packages. *J. Electron. Packag.* **2021**, *143*, 011001. [CrossRef]

13. Su, S.; Akkara, F.J.; Thaper, R.; Alkhazali, A.; Hamasha, M.; Hamasha, S. A state-of-the-art review of fatigue life prediction models for solder joint. *J. Electron. Packag.* **2019**, *141*, 040802. [CrossRef]
14. Lee, W.W.; Nguyen, L.T.; Selvaduray, G.S. Solder joint fatigue models: Review and applicability to chip scale packages. *Microelectron. Reliab.* **2000**, *40*, 231–244. [CrossRef]
15. Wong, E.H.; Vandriel, W.D.; Dasgupta, A.; Pecht, M. Creep fatigue models of solder joints: A critical review. *Microelectron. Reliab.* **2016**, *59*, 1–12. [CrossRef]
16. Tu, P.; Chan, Y.; Hung, K.; Lai, J. Growth kinetics of intermetallic compounds in chip scale package solder joint. *Scr. Mater.* **2001**, *44*, 317–323. [CrossRef]
17. Tian, Y.; Ren, N.; Jian, X.; Geng, T.; Wu, Y. Interfacial compounds characteristic and its reliability effects on SAC305 microjoints in flip chip assemblies. *J. Electron. Packag.* **2018**, *140*, 031007. [CrossRef]
18. Xian, J.W.; Belyakov, S.A.; Ivier, M.; Nogita, K.; Yasuda, H.; Gourlay, C.M. Cu_6Sn_5 crystal growth mechanisms during solidification of electronic interconnections. *Acta Mater.* **2017**, *126*, 540–551. [CrossRef]
19. Hwang, C.; Lee, J.; Suganuma, K.; Mori, H. Interfacial microstructure between Sn-3Ag-xBi alloy and Cu substrate with or without electrolytic Ni plating. *J. Electron. Mater.* **2003**, *32*, 52–62. [CrossRef]
20. Tu, K.N.; Lee, T.Y.; Jang, J.W.; Li, L.; Frear, D.R. Wetting reaction vs. solid-state aging of eutectic SnPb on Cu. *Appl. Phys.* **2001**, *89*, 4843–4849. [CrossRef]
21. Zeng, G.; Mcdonald, S.D.; Read, J.J.; Gu, Q.; Nogita, K. Kinetics of the polymorphic phase transformation of Cu_6Sn_5. *Acta Mater.* **2014**, *69*, 135–148. [CrossRef]
22. Zeng, K.; Tu, K.N. Six Cases of Reliability Study of Pb-free Solder Joints in Electronic Packaging Technology. *Mater. Sci. Eng.* **2002**, *38*, 55–105. [CrossRef]
23. Liu, B.; Tian, Y.; Qin, J.; An, R.; Zhang, R.; Wang, C. Degradation behaviors of micro ball grid array (mu BGA) solder joints under the coupled effects of electromigration and thermal stress. *J. Mater. Sci.-Mater. El.* **2016**, *27*, 11583–11592. [CrossRef]
24. Bashir, M.N.; Butt, S.U.; Mansoor, M.A.; Khan, N.B.; Bashir, S.; Wong, Y.H.; Alamro, T.; Eldin, S.M.; Jameel, M. Role of crystallographic orientation of beta-Sn grain on electromigration failures in lead-free solder joint: An overview. *Coatings* **2022**, *12*, 1752. [CrossRef]
25. Abdelhadi, O.; Ladani, L. IMC growth of Sn-3.5Ag/Cu system: Combined chemical reaction and diffusion mechanisms. *J. Alloys Compd.* **2012**, *537*, 87–99. [CrossRef]
26. Wang, Y.; Peng, X.; Huang, J.; Ye, Z.; Yang, J.; Chen, S. Studies of Cu-Sn interdiffusion coefficients in Cu_3Sn and Cu_6Sn_5 based on the growth kinetics. *Scr. Mater.* **2021**, *204*, 114138. [CrossRef]
27. Ma, H.R.; Dong, C.; Priyanka, P.; Wang, Y.P.; Li, X.G.; Ma, H.T.; Chen, J. Continuous growth and coarsening mechanism of the orientation-preferred Cu_6Sn_5 at Sn-3.0Ag/(001)Cu interface. *Mater. Charact.* **2020**, *166*, 110449. [CrossRef]
28. Ma, H.R.; Kunwar, A.; Shang, S.Y.; Jiang, C.R.; Wang, Y.P.; Ma, H.T.; Zhao, N. Evolution behavior and growth kinetics of intermetallic compounds at Sn/Cu interface during multiple reflows. *Intermetallics* **2018**, *96*, 1–12. [CrossRef]
29. Leineweber, A.; Wieser, C.; Hügel, W. Cu6Sn5 intermetallic: Reconciling composition and crystal structure. *Scr. Mater.* **2020**, *183*, 66–70. [CrossRef]
30. He, H.; Liu, X.; Wang, Z.; Hu, Q.; An, N.; Zhu, J.; Zhang, F.; Wang, L. Microstructural evolution of the Sn-51Bi-0.9Sb-1.0Ag/Cu soldering interface during isothermal aging. *J. Mater. Sci.-Mater. Electron.* **2021**, *32*, 15003–15010. [CrossRef]
31. Kunwar, A.; Shang, S.; Raback, P.; Wang, Y.; Givernaud, J.; Chen, J.; Ma, H.; Song, X.; Zhao, N. Heat and mass transfer effects of laser soldering on growth behavior of interfacial intermetallic compounds in Sn/Cu and Sn-3.5Ag0.5/Cu joints. *Microelectron. Reliab.* **2018**, *80*, 55–67. [CrossRef]
32. Mehreen, S.U.; Nogita, K.; Mcdonald, S.D.; Yasuda, H.; Stjohn, D.H. Peritectic phase formation kinetics of directionally solidifying Sn-Cu alloys within a broad growth rate regime. *Acta Mater.* **2021**, *220*, 117295. [CrossRef]
33. Wang, K.; Gan, D.; Hsieh, K. The orientation relationships of the Cu3Sn/Cu interfaces and a discussion of the formation sequence of Cu_3Sn and Cu_6Sn_5. *Thin Solid Film.* **2014**, *562*, 398–404. [CrossRef]
34. Cong, S.; Zhang, W.; Zhang, H.; Liu, P.; Tan, X.; Wu, L.; An, R.; Tian, Y. Growth kinetics of (CuxNi1-x)(6)Sn-5 intermetallic compound at the interface of mixed Sn63Pb37/SAC305 BGA solder joints during thermal aging test. *Mater. Res. Express* **2021**, *8*, 106301. [CrossRef]

Disclaimer/Publisher's Note: The statements, opinions and data contained in all publications are solely those of the individual author(s) and contributor(s) and not of MDPI and/or the editor(s). MDPI and/or the editor(s) disclaim responsibility for any injury to people or property resulting from any ideas, methods, instructions or products referred to in the content.

Article

Electronic and Electrical Properties of Island-Type Hybrid Structures Based on Bi-Layer Graphene and Chiral Nanotubes: Predictive Analysis by Quantum Simulation Methods

Michael M. Slepchenkov [1], Pavel V. Barkov [1] and Olga E. Glukhova [1,2,*]

[1] Institute of Physics, Saratov State University, Astrakhanskaya Street 83, 410012 Saratov, Russia
[2] Laboratory of Biomedical Nanotechnology, I.M. Sechenov First Moscow State Medical University, Trubetskaya Street 8-2, 119991 Moscow, Russia
* Correspondence: glukhovaoe@info.sgu.ru; Tel.: +7-8452-514562

Abstract: Hybrid structures based on graphene and carbon nanotubes (CNTs) are one of the most relevant modern nanomaterials for applications in various fields, including electronics. The variety of topological architectures of graphene/CNT hybrids requires a preliminary study of their physical properties by in silico methods. This paper is devoted to the study of the electronic and electrical properties of graphene/CNT hybrid 2D structures with an island topology using the self-consistent charge density functional-based tight-binding (SCC-DFTB) formalism and the Landauer–Buttiker formalism. The island-type topology is understood as the atomic configuration of a graphene/CNT hybrid film, in which the structural fragments of graphene and nanotubes form "islands" (regions of the atomic structure) with an increased density of carbon atoms. The island-type graphene/CNT hybrid structures are formed by AB-stacked bilayer graphene and (6,3)/(12,8) chiral single-walled carbon nanotubes (SWCNT). The bilayer graphene is located above the nanotube perpendicular to its axis. Based on the binding energy calculations, it is found that the atomistic models of the studied graphene/SWCNT hybrid structures are thermodynamically stable. The peculiarities of the band structure of graphene/SWCNT (6,3) and graphene/SWCNT (12,8) hybrid structures are analyzed. It is shown that the electronic properties of graphene/SWCNT hybrid structures are sensitive to the orientation and size of the graphene layers with respect to the nanotube surface. It is found that an energy gap of ~0.1 eV opens in the band structure of only the graphene/SWCNT (6,3) hybrid structure, in which the graphene layers of the same length are arranged horizontally above the nanotube surface. We revealed the electrical conductivity anisotropy for all considered atomistic models of the graphene/SWCNT (12,8) hybrid structure when bilayer graphene sheets with different sizes along the zigzag and armchair directions are located at an angle with respect to the nanotube surface. The obtained knowledge is important to evaluate the prospects for the potential application of the considered atomic configurations of graphene/SWCNT hybrid structures with island-type topology as connecting conductors and electrodes in electronic devices.

Keywords: graphene/carbon nanotube hybrid films; AB-stacked bilayer graphene; chiral single-walled carbon nanotubes; island topology; density functional tight binding method; electronic properties; band structure; energy gap; electrical conductivity anisotropy; Landauer–Buttiker formalism

1. Introduction

For several decades, graphene and carbon nanotubes have been the most discussed representatives of the carbon allotropes family [1–5]. Combining graphene and CNTs into a hybrid structure marked the transition to a new direction in materials science [6–15]. Graphene/CNT hybrid nanostructures have a larger surface area, porosity, thermal conductivity, mechanical strength, and improved optical, electrical, and electrochemical properties compared to their structural components [16–19]. Another advantage of graphene/CNT hybrid structures is improved hydrophobic characteristics. This makes them a promising material in environmental applications [20]. The excellent physical and chemical properties of graphene/CNT hybrid nanostructures open up wide opportunities for their application as flexible and transparent electrodes in field-effect transistors, energy storage devices, field emitters, sensors, and hydrogen storage systems [21–26].

Various approaches have been developed for the synthesis of graphene/CNT hybrid structures. They are usually divided into two groups, namely assembly methods and in situ methods [12]. Assembly methods use technologies such as vacuum filtration, sol-gel synthesis, layer-by-layer assembly, electrophoretic deposition, and solution processing [6]. Using the assembly methods, graphene/CNT hybrid structures with non-covalent interaction between CNTs and graphene are experimentally obtained. To obtain graphene/CNT hybrids with a covalent seamless junction of nanotubes and graphene, in situ, methods are used, including chemical vapor deposition, chemical unzipping, etc. [6]. The possibilities of modern synthesis technologies make it possible to obtain graphene/CNT hybrid structures with various architectures. Depending on the orientation of graphene and CNTs, all graphene/CNT hybrid structures are divided into three types [12]: (1) hybrids, where CNTs are horizontally oriented with respect to graphene; (2) hybrids, where CNTs are vertically oriented with respect to graphene; (3) hybrids, where CNTs are wrapped with graphene. The most common type of graphene/CNT hybrids is the first type listed above [12].

Computer simulation methods are actively used to conduct research aimed at revealing new physical effects and phenomena in graphene/CNT hybrid structures of various topologies. Quite a few papers are devoted to the theoretical study of the electronic and heat-conducting properties of seamless graphene/CNT heterostructures with vertically oriented CNTs [27–35]. Using the tight binding method, Matsumoto and Saito found that the graphene/SWCNT (6,6) hybrid structures have a band gap of 0.27 eV (in the case of SWCNTs with open ends) and 0.51 eV (in the case of SWCNTs with closed ends) [30]. A similar effect of opening a band gap of 0.2 eV was observed for a 3D carbon network formed by (5,5) SWCNTs embedded in graphene on both sides of the sheet [31]. Novaes et al. studied the transport properties of graphene/SWCNT hybrid structures with vertically oriented (4,4) and (8,0) SWCNTs using ab initio methods [32]. In some papers [33–35], the thermal conductivity of graphene/SWCNT hybrid structures with vertically oriented (6,6) SWCNTs was calculated. It has been shown that the heat flux in seamless 3D graphene/SWCNT (6,6) hybrid structures are determined by the minimum distance between the nanotubes and their length [34]. Interesting calculation results have been obtained for graphene/SWCNT hybrid structures with horizontally oriented SWCNTs [36–42]. The mechanical and electronic properties of graphene/SWCNT (12,0) and graphene/SWCNT (8,0) hybrids have been studied [36–38]. In these hybrid structures, nanotubes are covalently bonded to one or more graphene nanoribbons with a width equal to the SWCNT length. It has been established that the graphene/SWCNT (12,0) and graphene/SWCNT (8,0) hybrid structures are characterized by Van Hove singularities and a higher Young's modulus compared to individual nanotubes and graphene. In addition, the graphene/SWCNT (8,0) hybrid structure is characterized by the presence of an energy gap between the valence and conduction bands of several hundreds of a meV. Of great interest to researchers is quantum transport in graphene/SWCNT hybrid structures with horizontally oriented SWCNTs. The influence of such factors as the conductivity of SWCNTs [39], the graphene-SWCNT distance [40], and graphene nanoribbon's width and shape [41] on the quantum electron transport in graphene/SWCNT hybrid structures has been analyzed. The regularities of interaction of graphene/SWCNT hybrid 2D structures with covalently bonded graphene and horizontally oriented (10,0) and (12,0) SWCNTs with electromagnetic radiation in infrared, visible, and ultraviolet ranges have been studied [42]. It was shown that high-intensity optical conductivity peaks appear in the UV and optical ranges, regardless of the nanotube diameter and distance between them.

In the past few years, the attention of researchers has been turned to the development of technologies for obtaining graphene/SWCNT hybrid 2D structures with improved strength properties. In particular, Advincula et al. demonstrated the synthesis of a 2D graphene/SWCNT network by flash Joule heating of initial carbon nanotubes without the use of solvents and gases [43]. It was shown that a hybrid 2D network of covalently bonded SWCNTs and graphene is promising as an effective reinforcing additive in epoxy composites. The hardness and Young's modulus of graphene/SWCNT hybrid-based epoxy composites increase by 162% and 64%, respectively, compared with the neat epoxy. Li et al. have proposed an economical and scalable approach to obtain thin hybrid films based on SWCNTs and graphene nanoplatelets prepared by the spray-coating method [44]. In the hybrid structure, SWCNTs interacted with graphene nanoplates by means of van der Waals forces. In this case, the nanoplates overlapped each other and onto the nanotubes, forming networks with "islands" of increased carbon density. A strain sensor based on a hybrid graphene/SWCNT network is characterized by high sensitivity (calibration factor ~197 at 10% strain) and extensibility ($\geq$50%), as well as a reproducible response over 1000 load cycles. In addition, new prototypes of nanoelectronic devices based on graphene/SWCNT hybrid structures continue to be developed. Shin et al. have fabricated a p-type barrister based on a hybrid 2D structure formed by van der Waals joined graphene and a semiconductor SWCNT with a diameter of 1.3 nm [45]. The fabricated barrister showed an electron mobility of ~5350 cm^2/Vs and current on/off ratio of 10^6. Using palladium-catalyzed partial unzipping of SWCNTs, intramolecular heterojunctions based on graphene nanoribbons (2.4 nm in width) and SWCNTs (0.8 nm in diameter) were obtained [46]. The photovoltaic device based on these heterojunctions demonstrated a large open-circuit voltage of 0.52 V and a high efficiency of external power conversion of 4.7% under illumination with a wavelength of 1550 nm. Computational studies of recent years are mainly aimed at revealing the features of the atomic structure of various topological configurations of graphene/SWCNT hybrids and calculating their energy and electrical parameters. McDaniel conducted a fixed-voltage molecular dynamics study of the influence of structural features on the capacitance of electrodes based on a graphene/SWCNT hybrid [47]. The electrodes consist of different numbers of (9,9) or (12,12) SWCNTs stacked on a graphene sheet. It was found that the points of contact between graphene and SWCNTs serve as "hot spots" with significantly improved charge separation compared to the rest of the electrode. Xu and Jiang performed an MD study of the atomic configuration of a van der Waals heterostructure graphene/SWCNT/graphene using a mechanical model based on the competition between the bending energy and the adhesion energy [48]. According to the results of numerical simulations, it was found that the cross-section of the nanotube is compressed into an ellipse by graphene layers, and the eccentricity of the ellipse increases with increasing nanotube diameter. Wei and Zhang studied the formation mechanism of seamless junctions between vertically oriented SWCNTs (8,0) and a graphene monolayer with various topological defects [49] using the SCC-DFTB theory. Additionally, the authors calculated the Mülliken charge distribution for the construction of (8,0)-graphene SWCNT junctions. At the same time, in the above-mentioned papers with simulation results, graphene/SWCNT hybrid structures with nonchiral nanotubes are studied, while most of the synthesized SWCNTs are chiral nanotubes of sub-nanometer diameter [50].

The aim of this paper is to establish the effect of atomic structure features on the electronic properties and electrical conductivity of graphene/SWCNT hybrid structures with island-type topology formed by chiral nanotubes with a diameter of 0.6–1.3 nm and AB-stacked bilayer graphene. An island-type topology is understood as an atomic configuration of a graphene/SWCNT hybrid film in which the structural fragments of graphene and nanotubes form "islands" (regions of the atomic structure) with an increased density of carbon atoms). The features of the atomic structure will be understood as the mutual arrangement of graphene sheets in the bilayer graphene structure, the location of the bilayer graphene relative to the surface of the nanotube, the nanotube diameter, and the size of the graphene layer along the zigzag direction.

2. Calculation Details

The atomic structure and energy parameters of graphene/SWCNT hybrid structures with island topology were calculated using the SCC-DFTB method [51] implemented in the DFTB+ software package version 20.2 [52]. The tight binding approximation is included in the DFT model using perturbation theory [53]. This approximation is used at the stage of calculating the total energy of the system. The effect of electron density fluctuations on the total energy of the system is taken into account within the SCC-DFTB method. The distribution of the electron charge density over atoms is determined according to the Mulliken population analysis [54]. The valence approximation is used in the SCC-DFTB model. In accordance with the valence approximation, the largest contribution to the total energy of the system is made by the valence orbitals. In the course of SCC-DFTB calculations, the Slater-type orbitals with the set of pbc-0-3 parameters were used [52]. We chose the SCC-DFTB method due to the polyatomic nature of the supercells of graphene/SWCNT hybrid structures under study.

The electrical conductivity G was calculated within the framework of the Landauer–Buttiker formalism [55] according to the formula

$$G = \frac{I}{V} = \frac{2e^2}{h} \int_{-\infty}^{\infty} T(E) F_T(E - E_F) dE, \quad (1)$$

where $T(E)$ is the transmission function of electrons, F_T is the function of the thermal broadening of energy levels, E_F is the Fermi level of the electrodes, e is the elementary charge, h is Planck's constant, $2e^2/h$ is the doubled value of the conductance quantum to account for the spin. The $T(E)$ function is calculated using the equation [56]:

$$T(E) = \frac{1}{N} \sum_{k=1}^{N} \text{Tr}\left(\Gamma_S(E) G_C^A(E) \Gamma_D(E) G_C^R(E)\right), \quad (2)$$

where $G_C^R(E)$, $G_C^A(E)$ are the retarded and advanced Green's functions describing the contact with electrodes, $\Gamma_S(E)$, $\Gamma_D(E)$ are the level broadening matrices for the left (source) and drain (right) electrodes, respectively. Figure 1 shows the scheme for the calculation of quantum electron transport using one of the atomistic models of the graphene/SWCNT (6,3) hybrid 2D structure. The left and right electrodes and the conducting channel (scattering region) are supercells of the graphene/SWCNT hybrid structure. We considered the current transfer in the directions of the X (Figure 1a) and Y (Figure 1b) axes. In the case of the current transfer along the X axis, the electrodes are semi-infinite in the direction of X axis and infinite in the direction of Y axis. In the case of the current transfer along the Y axis, the electrodes are semi-infinite in the direction of Y axis and infinite in the direction of the X axis.

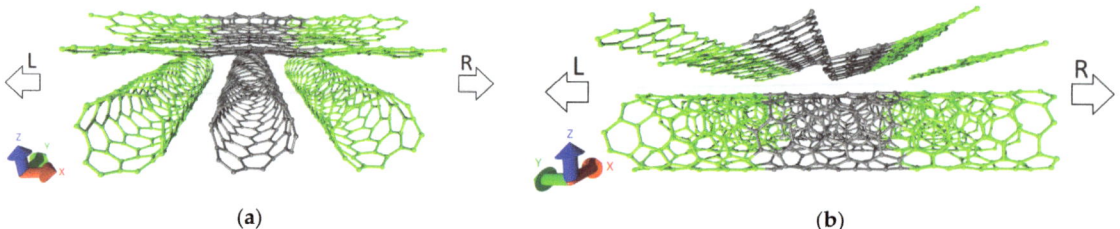

(a) (b)

Figure 1. The scheme for calculating the quantum electron transport in graphene/SWCNT (6,3) hybrid 2D structures for current transfer in the direction of X (**a**) and Y (**b**) axis. The left (it is marked with the letter L) and right (it is marked with the letter R) electrodes are highlighted in green. The scattering area (conducting channel) is highlighted in gray.

An original method was used to speed up the calculation of $T(E)$ of polyatomic supercells of graphene/SWCNT hybrids under study [57]. Within this method, the function

$T(E)$ was calculated for a small number of k-points of the first Brillouin zone. Then it was interpolated for any k-point of the first Brillouin zone and reconstructed the full function $T(E)$. The temperature of 300 K was used in all calculations.

3. Results and Discussion

3.1. Atomistic Models of Graphene/SWCNT Hybrid Structures with Island-Type Topology

Two types of configurations of graphene/SWCNT hybrid 2D structures with island-type topology are considered in this paper: (1) configurations based on (6,3) SWCNTs with a diameter of ~0.63 nm with a metallic type of conductivity; (2) configurations based on (12,8) SWCNTs with a diameter of ~1.3 nm with a semiconductor type of conductivity. For each topological configuration, three atomistic models of a supercell were constructed. The supercell contains fragment of AB-stacked bilayer graphene which is located above the nanotube surface, forming the so-called "islands" of increased density of carbon atoms. The structures with such topology are obtained during the synthesis of graphene/CNT hybrids with horizontally oriented nanotubes [44]. The atomistic models differed in the size of the graphene fragment along the Y axis (in the armchair direction of the graphene sheet) and in the value of the shift of one graphene layer relative to another along the Y axis. For (6,3) SWCNTs, atomistic models were constructed with a size of graphene fragment along the Y axis of 2 hexagons (model V1), three hexagons (model V2), and four hexagons (model V3). The shift of one layer of graphene relative to another along the Y axis was 0.069 nm for model V1, 0.375 nm for model V2, and 0.801 nm for model V3. Bilayer graphene and nanotube were located at a distance of ~0.3 nm for all three models. The layers in bilayer graphene were located at a distance of ~0.3 nm in the direction of the Z axis. The translation vectors of the supercells of models V1, V2 and V3 were L_x = 0.984 nm and L_y = 1.127 nm after optimization of the atomic structure. Supercells of models V1, V2, and V3 of graphene/SWCNT (6,3) hybrid 2D structures before and after optimization of the atomic structure are shown in Figure 2.

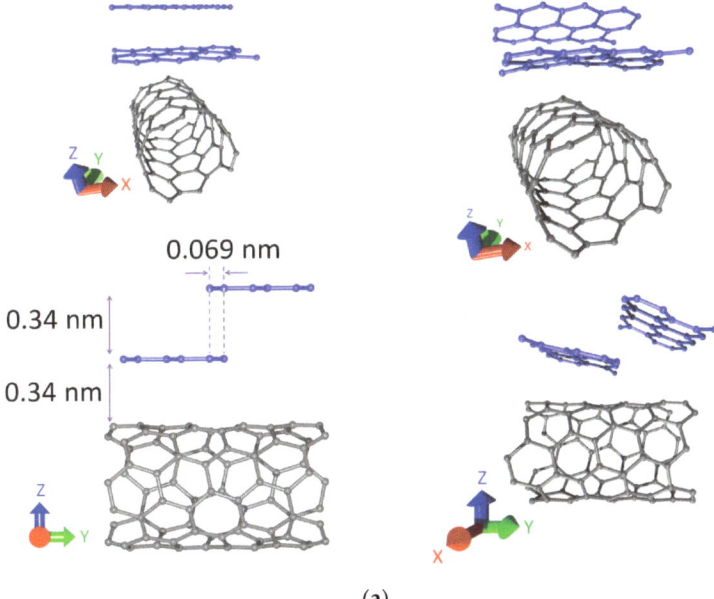

(a)

Figure 2. Cont.

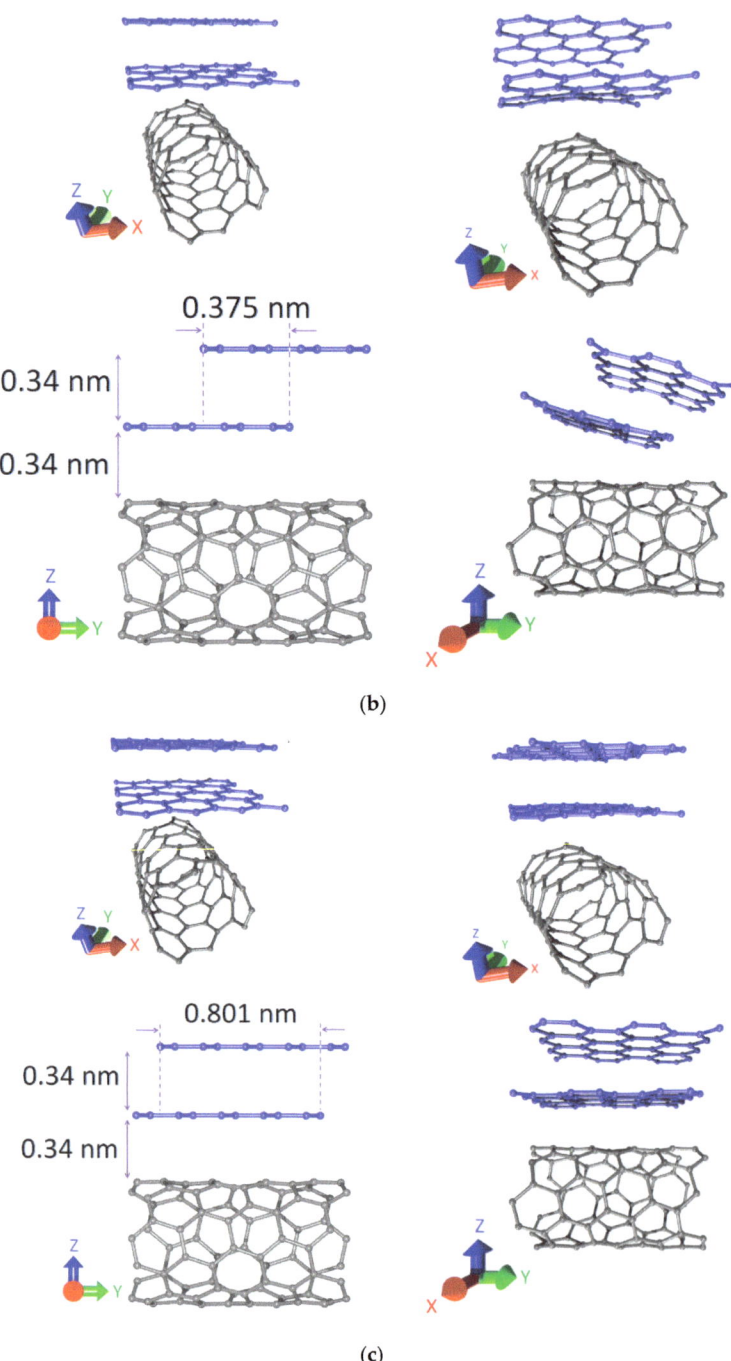

Figure 2. Supercells of graphene/SWCNT (6,3) hybrid 2D structures with island-type topology: (**a**) model V1; (**b**) model V2; (**c**) model V3.

As can be seen, in the cases of models V1 and V2, the sheets of bilayer graphene were deformed as a result of optimization of the atomic structure. They are located at a certain angle with respect to the nanotube surface. For model V3, the degree of deformation of bilayer graphene was minimal. The graphene sheets were arranged horizontally with respect to the nanotube surface. Differences in the orientation of graphene sheets with respect to the nanotube surface between the models are explained by the small size of graphene sheets in the armchair direction (along the Y axis) in the case of models V1 and V2. The nanotube (6,3) was almost not deformed due to its small size.

For (12,8) SWCNTs, atomistic models were constructed with a size of graphene fragment along the Y axis of 4 hexagons (model V1), five hexagons (models V2), and six hexagons (model V3). Since the length of SWCNT (1.857 nm) in the supercell of the graphene/SWCNT (12,8) hybrid structure is noticeably larger than in the supercell of the graphene/SWCNT (6,3) hybrid structure (1.127 nm), the size of the graphene fragment along the Y axis for topological configurations with (12,8) SWCNTs was chosen larger. The shift of one layer of graphene relative to another along the Y axis was 0.131 nm for model V1, 0.557 nm for model V2, and 0.983 nm for model V3. The distance between the graphene layers along the Z axis, as well as the distance between the graphene bilayer and nanotube, was chosen to be ~0.3 nm for all three models. The translation vectors of the supercells of models V1, V2 and V3 after optimization of the atomic structure were the same and amounted to L_x = 1.857 nm and L_y = 1.968 nm. Supercells of models V1, V2, and V3 of the graphene/SWCNT (12,8) hybrid 2D structures before and after optimization of the atomic structure are shown in Figure 3. It is clearly seen that, for all three models, a noticeable deformation of graphene sheets and nanotubes in the supercell is observed. The nanotube acquires the shape of an ellipsoid, shrinking along the Z axis.

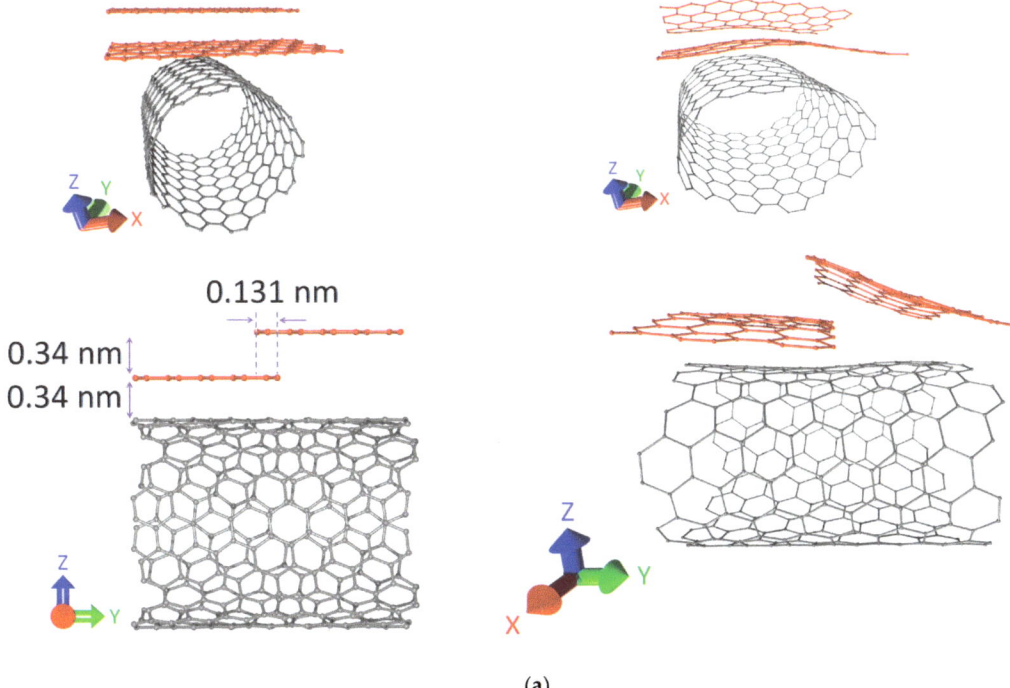

(a)

Figure 3. *Cont.*

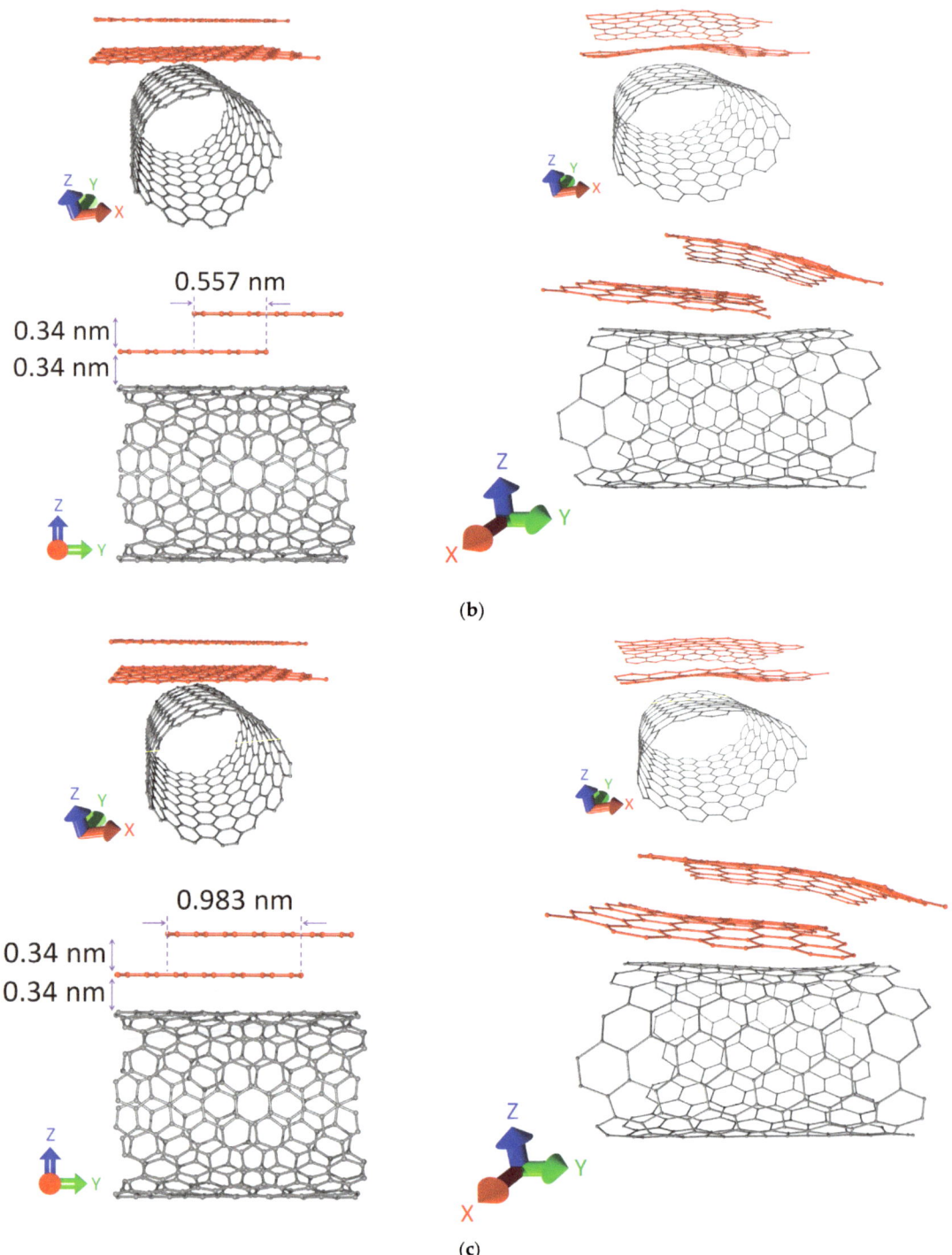

Figure 3. Supercells of bilayer graphene-SWCNT (12,8) hybrid 2D structures with island-type topology: (**a**) model V1; (**b**) model V2; (**c**) model V3.

The constructed supercells were tested for thermodynamic stability at room temperature. To do this, we calculated the binding energy E_b according to Equation (3):

$$E_b = (E_h - E_{gr} - E_{tube})/N, \qquad (3)$$

where E_h is the energy of a graphene/SWCNT hybrid structure, E_{gr} is the energy of a bilayer graphene, E_{tube} is the energy of a nanotube, and N is the number of atoms in a supercell. The calculated values of E_b are given in Table 1. According to Table 1, each of the considered configurations is characterized by a negative binding energy. Therefore, the resulting atomic configurations of the supercells are energetically favorable.

Table 1. Energy characteristics of the supercells of graphene/SWCNT (6,3) and graphene/SWCNT (12,8) hybrid 2D structures with island-type topology.

Characteristics	V1	V2	V3
	graphene/SWCNT (6,3) hybrid structures		
E_b, eV/atom	−0.012	−0.109	−0.133
	graphene/SWCNT (12,8) hybrid structures		
E_b, eV/atom	−0.017	−0.017	−0.016

Table 1 shows that model V3 has the highest thermodynamic stability among the considered supercells of the graphene/SWCNT (6,3) hybrid structures. This is due to the low degree of deformation of graphene sheets and the absence of deformation for the nanotube. The degree of deformation of the nanotube and graphene sheets in the supercells of the graphene/SWCNT (12,8) hybrid structures are approximately the same for all models. Therefore, the binding energies of these models are close in magnitude.

3.2. Electronic Properties of Graphene/SWCNT Hybrid Structures of Island Type

To understand the prospects for using the graphene/SWCNT (6,3) and graphene/SWCNT (12,8) hybrid films in nanoelectronic devices, it is necessary (1) to know the features of the electronic structure, including the patterns of formation of the band structure; (2) to understand the pattern of quantum electron transport, including for controlling the electrical conductivity. The key point of the study is to establish the possibility of topological control of the electronic–energetic and electrophysical parameters of graphene/SWCNT hybrid films. We consider the influence of such topological features as the mutual arrangement of graphene sheets in the bilayer graphene structure, the orientation of the bilayer graphene with respect to the nanotube surface, the nanotube diameter, and the size of the graphene layer along the zigzag direction.

To reveal the electronic structure features of graphene/SWCNT hybrid films with island-type topology, the energy band diagrams were calculated. The band structure was sampled along a path M–Γ–J–K–Γ within the first Brillouin zone. One of our tasks was to analyze what features of the electronic structure and properties of bilayer graphene and (6,3) and (12,8) SWCNTs are manifested in the graphene/SWCNT hybrid structures. Figures 4–6 show the calculated band diagrams near the Fermi level for supercells of graphene/SWCNT (6,3) hybrid structures. These figures also show the band diagrams of a nanotube (6,3) and bilayer graphene, which are part of the supercell of the graphene/SWCNT (6,3) hybrid structure. Based on the results of the analysis of the presented diagrams, the following patterns can be identified. The band diagrams of graphene/SWCNT (6,3) hybrid structures of all three models contain characteristic features of the electronic structure of both bilayer graphene and (6,3) SWCNTs. The contribution of bilayer graphene to the energy profile near the valence band maximum (VBM) and near the conduction band minimum (CBM) is clearly seen between the highly symmetrical J and K points of the Brillouin zone. The contribution of (6,3) SWCNTs to the energy profile near the VBM and CBM manifests itself between Γ and J points. Between M and Γ, as well as between K and Γ points, the joint influence of bilayer graphene and (6,3) SWCNTs on the patterns of the energy profile of the graphene/SWCNT (6,3) hybrid structure.

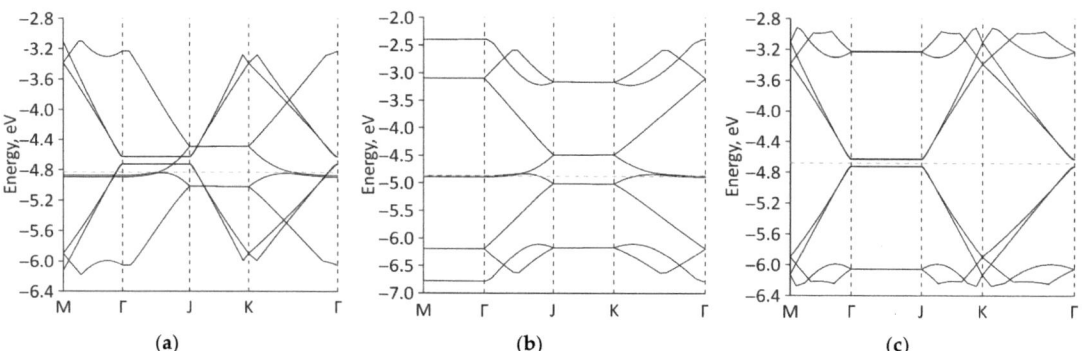

Figure 4. Band diagrams near the Fermi level of the model V1 of graphene/SWCNT (6,3) hybrid structures with island-type topology: (**a**) hybrid structure; (**b**) bilayer graphene; (**c**) SWCNT (6,3). The green dotted line shows the Fermi level.

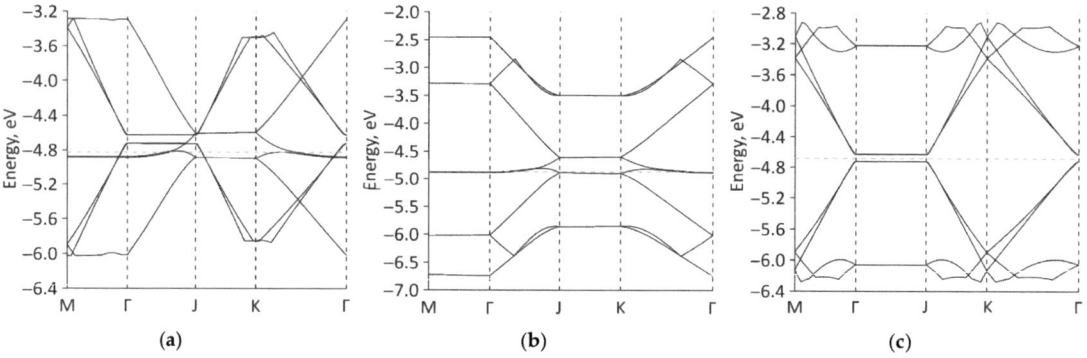

Figure 5. Band diagrams near the Fermi level of the model V2 of graphene/SWCNT (6,3) hybrid structures with island-type topology: (**a**) hybrid structure; (**b**) bilayer graphene; (**c**) SWCNT (6,3). The green dotted line shows the Fermi level.

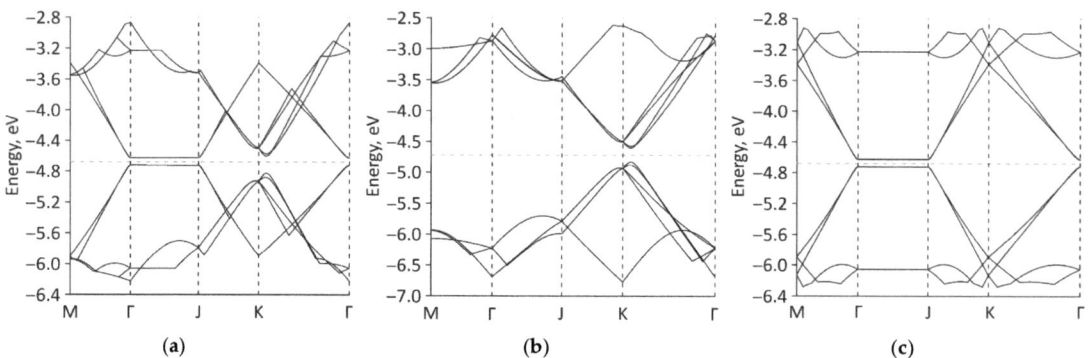

Figure 6. Band diagrams near the Fermi level of the model V3 of graphene/SWCNT (6,3) hybrid structures with island-type topology: (**a**) hybrid structure; (**b**) bilayer graphene; (**c**) SWCNT (6,3). The green dotted line shows the Fermi level.

An analysis of the band diagrams makes it possible to trace how the type of conductivity of the graphene/SWCNT (6,3) hybrid structure changes depending on the topological features of the supercells of models V1, V2, and V3. For supercells of models V1 and V2, no energy gap is observed between the VBM and CBM, which indicates the gapless nature of the band structure of these atomic configurations of graphene/SWCNT hybrids (6,3). A similar picture is also characteristic of the band structure of bilayer graphene, which is part of the supercells of these models. Model V3 is characterized by the appearance of an energy gap of ~0.1 eV between the VBM and CBM, which allows us to speak about the opening of an energy gap in the band structure. To explain this physical phenomenon, we considered how the position of the Fermi level changes in the energy band diagrams of the graphene/SWCNT (6,3) hybrid structure and its components (bilayer graphene and (6,3) SWCNTs). The values of the Fermi level are presented in Table 2. It can be seen that that in the case of models V1 and V2, the Fermi level (−4.82 eV) of the graphene/SWCNT (6,3) hybrid structure almost completely repeats the position of the Fermi level of the bilayer graphene fragment in the supercells of these topological models (−4.87 eV for the model V1and −4.86 eV for the model V2). As is known, it is the position of the Fermi level that determines both the type of conductivity and the basic electrophysical properties of the material. Taking into account the gapless band structure of a fragment of bilayer graphene from supercells of models V1 and V2 (see Figures 4 and 5), we can say that for these models, it is bilayer graphene that makes a decisive contribution to the electronic properties and type of conductivity of the graphene/SWCNT (6,3) hybrid structure. In the case of model V3, the Fermi level (−4.67 eV) of the graphene/SWCNT (6,3) hybrid structure completely coincides with the Fermi level of the SWCNT (6,3) from the supercell of model V3. The size of the opened energy gap in the band structure of the graphene/SWCNT (6,3) hybrid is the same as that of the SWCNT (6,3). In addition, the energy gap in the band structure of the graphene/SWCNT (6,3) hybrid opens between the highly symmetrical points Γ and J of the Brillouin zone, just as in SWCNT (6,3). At the same time, in the case of model V3, an energy gap also opens in the band structure of a bilayer graphene fragment from the graphene/SWCNT (6,3) supercell. The Fermi level (−4.71 eV) of the bilayer graphene fragment from the model V3 supercell changes significantly compared to the Fermi level of the bilayer graphene fragment from the V1 and V2 models (−4.82 eV) and approaches the Fermi level of the hybrid structure (−4.67 eV). All this allows us to assume that in the case of model V3, both bilayer graphene and (6,3) SWCNTs affect the appearance of an energy gap in the band structure of the graphene/SWCNT (6,3) hybrid, but the contribution of (6,3) SWCNTs is decisive.

Table 2. Fermi level of graphene/SWCNT (6,3) and graphene/SWCNT (12,8) hybrid structures and their individual components.

Atomistic Model	Graphene/SWCNT	Bilayer Graphene	SWCNT
graphene/SWCNT (6,3) hybrid structures			
model V1	−4.828	−4.873	−4.674
model V2	−4.822	−4.867	−4.674
model V3	−4.674	−4.714	−4.674
graphene/SWCNT (12,8) hybrid structures			
model V1	−4.858	−4.858	−4.683
model V2	−4.849	−4.849	−4.683
model V3	−4.840	−4.840	−4.682

The reason for such noticeable differences between the model V3 and the models V1 and V2 is the different topology of bilayer graphene in these models. Let us illustrate them visually by the example of extended fragments of each of the models obtained by multiple translations of their supercells in two directions (along the X and Y axis). These fragments are shown in Figure 7. It can be seen that, for models V1 and V2, graphene layers of small width in the direction of the Y axis line up one after another at an angle with respect to the nanotube surface. In this case, the edge atoms of one graphene layer are located above the edge atoms of the other graphene layer, causing the curvature of their atomic network. In the extended fragment of model V3, the graphene layers are oriented horizontally with respect to the nanotube surface and have similar sizes in both directions of the supercell translation. Due to the arrangement of graphene layers of the same size in a horizontal plane, their atomic network has a minimum curvature, and the interaction at the level of electron orbitals is much weaker as a result. In addition, Figure 7c shows that the upper graphene layer in model V3 slightly rotated relative to the lower one, forming a twisted graphene structure. As was previously found, twisted bilayer graphene demonstrates the effect of opening the energy gap between the valence band and the conduction band [58].

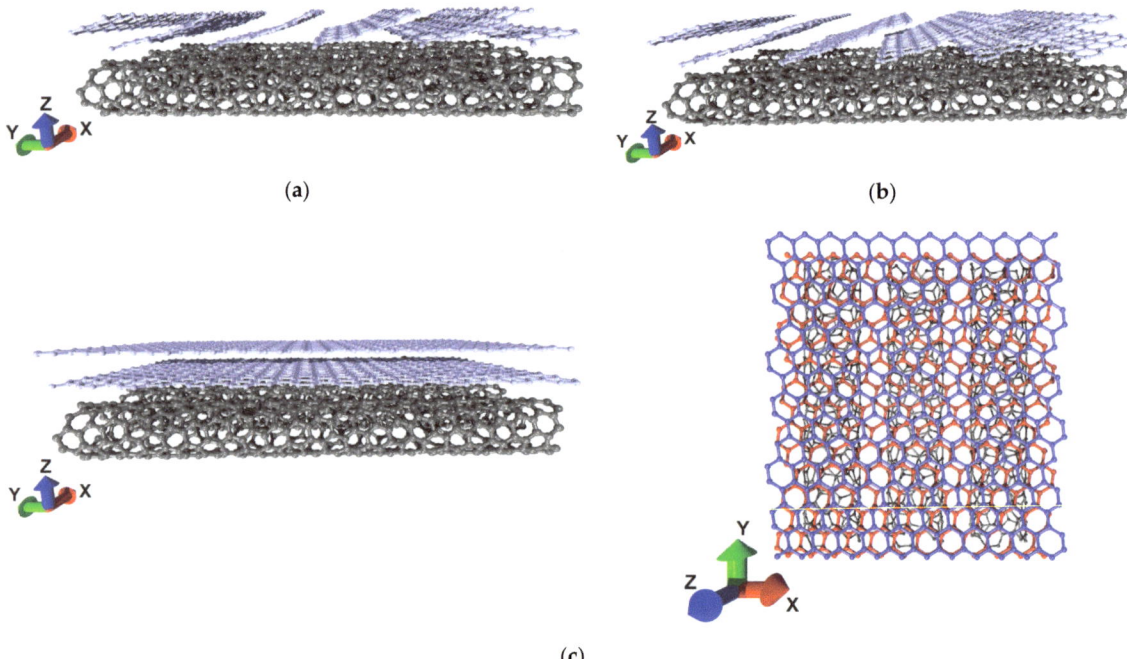

Figure 7. Extended fragments of the supercells of graphene/SWCNT (6,3) hybrid structures with island-type topology: (**a**) model V1; (**b**) model V2; (**c**) model V3 (profile and top view).

The analysis of the electronic structure for the topological models of the graphene/SWCNT (12,8) hybrid structure was carried out in a similar manner. Let us trace the pattern of the energy subband profile of the graphene/SWCNT (12,8) hybrid structure using the model V1 as an example. Figure 8 shows the energy band diagrams of model V1 of the graphene/SWCNT (12,8) hybrid structure and its individual structural components. It can be seen that the graphene/SWCNT (12,8) hybrid has a gapless band structure, which is characteristic of bilayer graphene in the composition of the model V1 supercell. A completely identical picture of the energy subband profile of the band diagram is also observed for models V2 and V3. Since the degree of deformation of graphene layers and nanotube (12,8) is the same in the supercells of models V1, V2 and V3, no visible changes in the band structure of graphene/SWCNT (12,8) hybrid and its individual components are observed. Therefore, we confine ourselves to presenting the energy band diagrams for model V1.

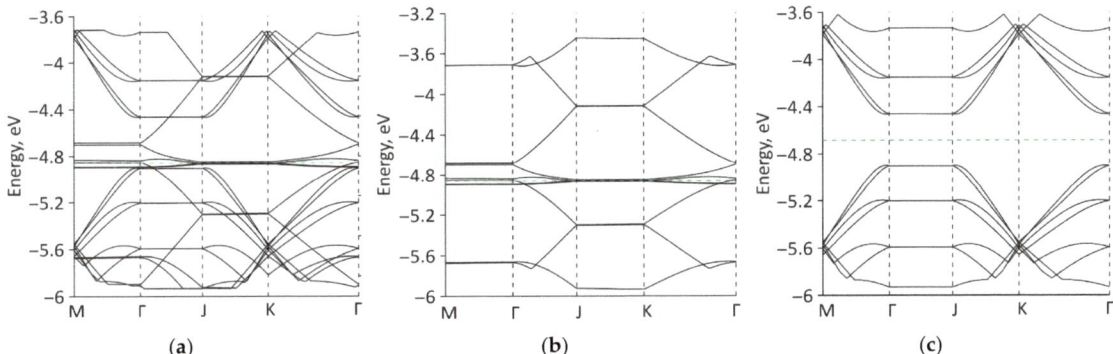

Figure 8. Band diagrams near the Fermi level of the model V1 of graphene/SWCNT (12,8) hybrid structure with island-type topology: (**a**) hybrid structure; (**b**) bilayer graphene; (**c**) SWCNT (12,8). The green dotted line shows the Fermi level.

The absence of differences in the band diagrams of the models V1, V2, and V3 of the graphene/SWCNT (12,8) hybrid structure can be explained by their similar topological features. This is clearly confirmed by the extended fragments of each of the models shown in Figure 9. In each of the cases, the graphene layers in the extended fragment are located at some angle with respect to the nanotube surface, like "steps of a ladder". A similar pattern was observed earlier for models V1 and V2 of the graphene/SWCNT (6,3) hybrid structure.

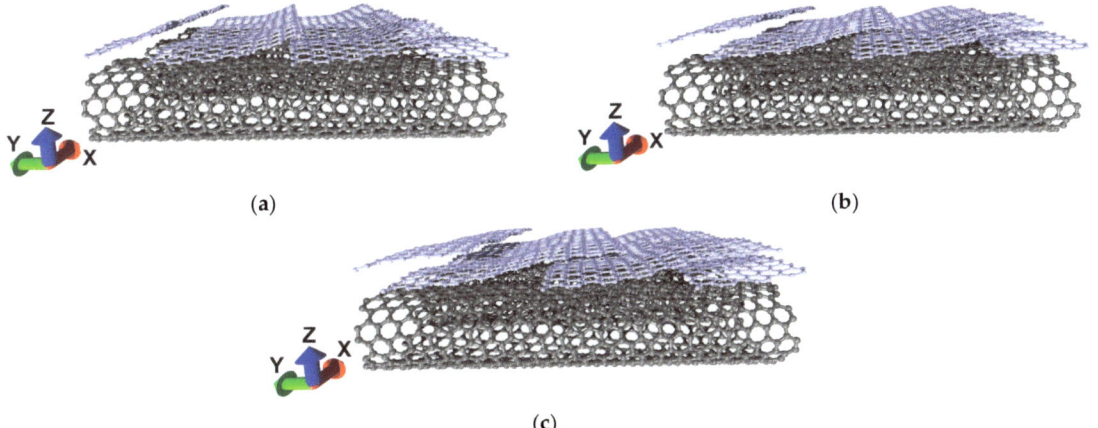

Figure 9. Extended fragments of the atomic structure of models of graphene/SWCNT (12,8) hybrid structures with an island topology: (**a**) model V1; (**b**) model V2; (**c**) model V3.

The leading role of bilayer graphene in determining the type of conductivity of the graphene/SWCNT (12,8) hybrid structure is confirmed by the complete coincidence of the location of the Fermi level in their band structures: −4.858 eV for the model V1, −4.849 eV for the model V2, and −4.840 for the model V3.

3.3. Electrical Properties of Graphene/SWCNT Hybrid Structures with Island-Type Topology

Having revealed the features of the electronic structure of the graphene/SWCNT (6,3) and graphene/SWCNT (12,8) hybrid structures with the island-type topology, we proceed to a discussion of their electrical properties. We evaluated the electrical properties by the magnitude of the electrical resistance, which is the main parameter of the connecting conductors in the circuits of various electronic devices. Table 3 shows the electrical resistances R_X and R_Y calculated for models V1, V2 and V3 of the graphene/SWCNT (6,3) and graphene/SWCNT (12,8) hybrid structures in two directions of current transfer: along the zigzag direction (X axis) and along the armchair direction (Y axis) of graphene hexagonal lattice. The electrical resistance was defined as the reciprocal of the electrical conductivity G calculated according to Equation (2).

Table 3. The electrical resistances of the supercells of graphene/SWCNT (6,3) and graphene/SWCNT (12,8) hybrid structures with island-type topology.

Characteristics	V1	V2	V3
	graphene/SWCNT (6,3) hybrid structures		
R_X, khOhm	7.068	5.942	126.287
R_Y, khOhm	6.125	6.066	12.215
	graphene/SWCNT (12,8) hybrid structures		
R_X, khOhm	6.808	6.419	5.949
R_Y, khOhm	38.080	55.414	100.162

Table 3 shows that the resistance values of models V1 and V2 of the graphene/SWCNT (6,3) hybrid structure are almost the same in both directions of current transfer. The similarity in the resistance values of the models V1 and V2 is explained by the similarity of their topological features discussed above. In the case of model V3, the difference in the resistance values between the current transfer directions is almost 10 times. This indicates the presence of electrical conductivity anisotropy. In order to explain the observed anisotropy, Figure 10 shows graphs of the transmission function $T(E)$ for model V3 (graphene/SWCNT (6,3) hybrid structure and its individual components) in two directions of current transfer (along the X and Y axes). It can be seen from the figure that the electrical conductivity anisotropy is due to the different number of conduction channels in the directions of current transfer: the values of $T(E)$ near the Fermi level at current transport along the X axis are several times larger than for current transport along the Y axis. Consequently, in the case of current transport along the X axis, the graphene/SWCNT (6,3) hybrid takes on the properties of bilayer graphene; in the case of current transport along the Y axis, the graphene/SWCNT (6,3) hybrid takes the properties of SWCNT (6,3).

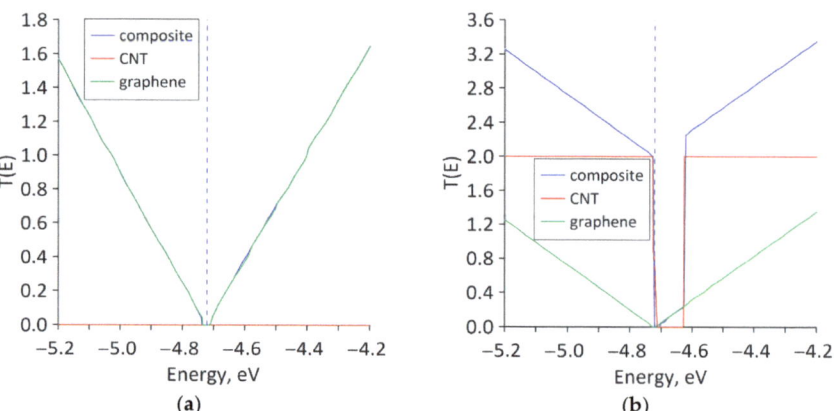

Figure 10. Transmission functions of the graphene/SWCNT (6,3) hybrid structure and its individual components for model V3: (**a**) current transport along the X axis; (**b**) current transport along the Y axis. The dotted vertical line marks the Fermi level of the graphene/SWCNT (6,3) hybrid structure.

In the case of the graphene/SWCNT (12,8) hybrid structure, the electrical conductivity anisotropy is observed for all three atomistic models. Let us explain the observed anisotropy using model V1 as an example, analyzing the calculated $T(E)$ profiles for the graphene/SWCNT (12,8) hybrid structure and its individual structural components. These profiles are shown in Figure 11. As in the case of the model V3 of the graphene/SWCNT (6,3) hybrid structure, the appearance of anisotropy is associated with a different number of conduction channels in two directions of current transfer: in the direction of the X axis, the values of $T(E)$ near the Fermi level are several times larger than in the direction of Y axis. With an increase in the size of bilayer graphene along the Y axis (in the direction of the armchair of the graphene hexagonal lattice), the overlap area of the graphene layers increases, and, hence, the intensity of the electronic interaction between them and the nanotube in the region of "islands" of increased density of carbon atoms. This can explain the increase in the resistance R_Y during the transition from model V1 to model V3.

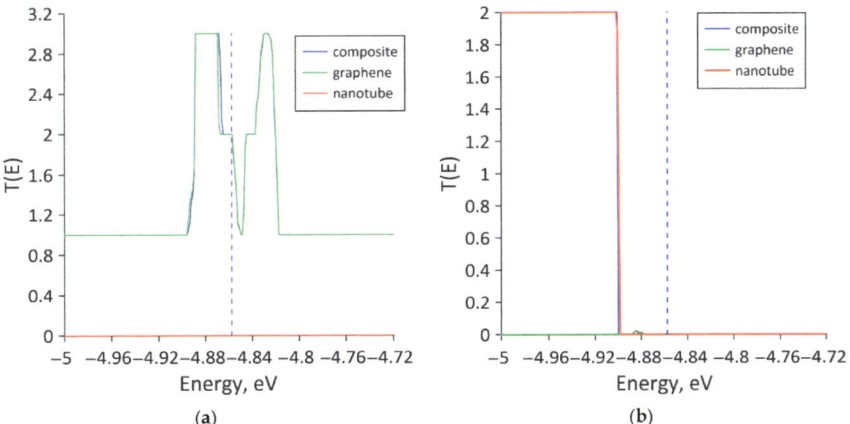

Figure 11. Transmission functions of the graphene/SWCNT (12,8) hybrid structure and its individual components for model V1: (**a**) current transfer along the X axis; (**b**) current transfer along the Y axis. The dotted vertical line marks the Fermi level of the graphene/SWCNT (12,8) hybrid structure.

The electrical conductivity anisotropy revealed for model V3 of the graphene/SWCNT (6,3) hybrid structure and models V1, V2, and V3 of the graphene/SWCNT (12,8) hybrid structure is related to the topological and electronic structure features of bilayer graphene in the supercells of these models. Based on the data in Table 3, it can be concluded that the models V1 and V2 of the graphene/SWCNT (6,3) hybrid structure has the lowest electrical resistance values (~6–7 kOhm) in both directions of current transport. In terms of electrical resistance values, these atomic configurations of graphene/SWCNT hybrid structures are not inferior to the structures of pillared graphene (~10–30 kOhm) [57], hybrid films based on oxidized graphene and MWCNTs (~20 kOhm) [58], some varieties of carbon nanotube-reinforced polymer composites (~10–20 kOhm) [59] and graphene-based fiber-reinforced composite (~200 kOhm) [60].

4. Conclusions

In this article, for the first time, the issues of topological control of the electronic and electrical properties of graphene-nanotube hybrid films with island-type topology (fragments of graphene and nanotubes form "islands" with an increased density of carbon atoms) were considered in detail by changing various topological and geometric parameters of the hybrid film: nanotube diameter, size of the graphene fragment along the zigzag direction (along the Y axis), the shift of one layer of graphene relative to another along the zigzag direction (along the Y axis), orientation (tilt angle) of the bilayer graphene with respect to the nanotube surface. To build atomistic models of graphene/SWCNT hybrid 2D structures, we chose chiral (6,3) and (12,8) SWCNTs of sub- and nanometer diameters, which are most often encountered in the experiment. Based on the results of energy band diagram calculations, it was established that the electronic properties of graphene/SWCNT hybrid structures, in particular, the type of conductivity, are sensitive to the orientation and size of the graphene layers with respect to the nanotube surface. It has been found that an energy gap of ~0.1 eV opens in the band structure of the graphene/SWCNT (6,3) hybrid when graphene layers of the same length are arranged horizontally above the nanotube surface. When bilayer graphene sheets with different sizes along the zigzag and armchair directions are located at an angle with respect to the nanotube surface, for the graphene/SWCNT (12,8) hybrid structure, electrical conductivity anisotropy is observed: the electrical resistance in the zigzag direction is smaller than in the direction of the graphene hexagonal lattice. The anisotropy is caused by a different number of conduction channels along the zigzag and armchair directions. The advantage of the zigzag direction is due to the topology of the bilayer graphene fragment in the supercells of the considered models of graphene/SWCNT (12,8) hybrid films. The graphene bilayer has the shape of a zigzag nanoribbon, which is known to be characterized by the presence of localized edge states with energies close to the Fermi level [61]. The smallest electrical resistance was ~6 kOhm for both graphene/SWCNT (6,3) and graphene/SWCNT (12,8) hybrid structures.

The obtained calculation results are of great importance for nanotechnologists and developers of electronic nanodevices. On the one hand, the results of the predictive modeling carried out make it possible to implement the concept of planned experiments to obtain graphene/SWCNT hybrid structures with desired properties. On the other hand, they make it possible to conclude that the considered graphene/SWCNT thin hybrid films with island-type topology have prospects for potential application as connecting conductors and electrodes in nanoelectronic devices.

Author Contributions: Conceptualization, O.E.G. and M.M.S.; methodology, O.E.G. and M.M.S.; funding acquisition, O.E.G., P.V.B. and M.M.S.; investigation, O.E.G., P.V.B. and M.M.S.; writing—original draft preparation, P.V.B.; writing—review and editing, O.E.G. and M.M.S. All authors have read and agreed to the published version of the manuscript.

Funding: The research was funded by the Ministry of Science and Higher Education of the Russian Federation (project No. FSRR-2023-0008).

Institutional Review Board Statement: Not applicable.

Informed Consent Statement: Not applicable.

Data Availability Statement: Not applicable.

Conflicts of Interest: The authors declare no conflict of interest.

References

1. Slepičková Kasálková, N.; Slepička, P.; Švorčík, V. Carbon Nanostructures, Nanolayers, and Their Composites. *Nanomaterials* **2021**, *11*, 2368. [CrossRef] [PubMed]
2. Urade, A.R.; Lahiri, I.; Suresh, K.S. Graphene Properties, Synthesis and Applications: A Review. *JOM* **2023**, *75*, 614–630. [CrossRef] [PubMed]
3. Rathinavel, S.; Priyadharshini, K.; Panda, D. A review on carbon nanotube: An overview of synthesis, properties, functionalization, characterization, and the application. *Mater. Sci. Eng. B* **2021**, *268*, 115095. [CrossRef]
4. Chen, Y.; Long, J.; Xie, B.; Kuang, Y.; Chen, X.; Hou, M.; Gao, J.; Liu, H.; He, Y.; Wong, C.P. One-Step Ultraviolet Laser-Induced Fluorine-Doped Graphene Achieving Superhydrophobic Properties and Its Application in Deicing. *ACS Appl. Mater. Interfaces* **2022**, *14*, 4647–4655. [CrossRef]
5. Wang, J.; Wang, N.; Xu, D.; Tang, L.; Sheng, B. Flexible humidity sensors composed with electrodes of laser induced graphene and sputtered sensitive films derived from poly(ether-ether-ketone). *Sens. Actuators B Chem.* **2023**, *375*, 132846. [CrossRef]
6. Wu, X.; Mu, F.; Zhao, H. Recent progress in the synthesis of graphene/CNT composites and the energy-related applications. *J. Mater. Sci. Technol.* **2020**, *55*, 16–34. [CrossRef]
7. Jomol, P.J.; Mary Nancy, T.E.; Bindu Sharmila, T.K. A comprehensive review on the environmental applications of graphene–carbon nanotube hybrids: Recent progress, challenges and prospects. *Mater. Adv.* **2021**, *2*, 6816–6838.

8. Lv, R.; Cruz-Silva, E.; Terrones, M. Building Complex Hybrid Carbon Architectures by Covalent Interconnections: Graphene-Nanotube Hybrids and More. *ACS Nano* **2014**, *8*, 4061–4069. [CrossRef]
9. Barshutina, M.N.; Volkov, V.S.; Arsenin, A.V.; Nasibulin, A.G.; Barshutin, S.N.; Tkachev, A.G. Silicone Composites with CNT/Graphene Hybrid Fillers: A Review. *Materials* **2021**, *14*, 2418. [CrossRef]
10. Liao, Y.; Mustonen, K.; Tulić, S.; Skákalová, V.; Khan, S.A.; Laiho, P.; Zhang, Q.; Li, C.; Monazam, M.R.A.; Kotakoski, J.; et al. Enhanced Tunneling in a Hybrid of Single-Walled Carbon Nanotubes and Graphene. *ACS Nano* **2019**, *13*, 11522–11529. [CrossRef]
11. Gorkina, A.L.; Tsapenko, A.P.; Gilshteyn, E.P.; Koltsova, T.S.; Larionova, T.V.; Talyzin, A.; Anisimov, A.S.; Anoshkin, I.V.; Kauppinen, E.I.; Tolochko, O.V.; et al. Transparent and conductive hybrid graphene/carbon nanotube films. *Carbon* **2016**, *100*, 501–507. [CrossRef]
12. Fan, W.; Longsheng, Z.; Tianxi, L. *Graphene-Carbon Nanotube Hybrids for Energy and Environmental Applications*, 1st ed.; Springer: Singapore, 2017; pp. 21–51.
13. Xia, K.; Zhan, H.; Gu, Y. Graphene and Carbon Nanotube Hybrid Structure: A Review. *Procedia IUTAM* **2017**, *21*, 94–101. [CrossRef]
14. Dang, V.T.; Nguyen, D.D.; Cao, T.T.; Le, P.H.; Tran, D.L.; Phan, N.M.; Nguyen, V.C. Recent trends in preparation and application of carbon nanotube–graphene hybrid thin films. *Adv. Nat. Sci. Nanosci. Nanotechnol.* **2016**, *7*, 033002. [CrossRef]
15. Dasgupta, A.; Rajukumar, L.P.; Rotella, C.; Lei, Y.; Terrones, M. Covalent three-dimensional networks of graphene and carbon nanotubes: Synthesis and environmental applications. *Nano Today* **2017**, *12*, 116–135. [CrossRef]
16. Kuang, J.; Dai, Z.; Liu, L.; Yang, Z.; Jinc, M.; Zhang, Z. Synergistic effects from graphene and carbon nanotubes endow ordered hierarchical structure foams with a combination of compressibility, super-elasticity and stability and potential application as pressure sensors. *Nanoscale* **2015**, *7*, 9252–9260. [CrossRef] [PubMed]
17. Kholmanov, I.N.; Magnuson, C.W.; Piner, R.; Kim, J.Y.; Aliev, A.E.; Tan, C.; Kim, T.Y.; Zakhidov, A.A.; Sberveglieri, G.; Baughman, R.H.; et al. Optical, electrical, and electromechanical properties of hybrid graphene/carbon nanotube films. *Adv. Mater.* **2015**, *27*, 3053–3059. [CrossRef] [PubMed]
18. Gan, X.; Lv, R.; Bai, J.; Zhang, Z.; Wei, J.; Huang, Z.H.; Zhu, H.; Kang, F.; Terrones, M. Efficient photovoltaic conversion of graphene–carbon nanotube hybrid films grown from solid precursors. *2D Mater.* **2015**, *2*, 034003. [CrossRef]
19. Maarouf, A.A.; Kasry, A.; Chandra, B.; Martyna, G.J. A graphene-carbon nanotube hybrid material for photovoltaic applications. *Carbon* **2016**, *102*, 74–80. [CrossRef]
20. Wan, W.; Zhang, R.; Li, W.; Liu, H.; Lin, Y.; Li, L.; Zhou, Y. Graphene–carbon nanotube aerogel as an ultra-light, compressible and recyclable highly efficient absorbent for oil and dyes. *Environ. Sci. Nano* **2016**, *3*, 74–80. [CrossRef]
21. Shi, E.; Li, H.; Yang, L.; Hou, J.; Li, Y.; Li, L.; Cao, A.; Fang, Y. Carbon nanotube network embroidered graphene films for monolithic all-carbon electronics. *Adv. Mater.* **2015**, *27*, 682–688. [CrossRef]
22. Kim, S.H.; Song, W.; Jung, M.W.; Kang, M.A.; Kim, K.; Chang, S.J.; Lee, S.S.; Lim, J.; Hwang, J.; Myung, S.; et al. Carbon Nanotube and Graphene Hybrid Thin Film for Transparent Electrodes and Field Effect Transistors. *Adv. Mater.* **2014**, *26*, 4247–4252. [CrossRef]
23. Pyo, S.; Eun, Y.; Sim, J.; Kim, K.; Choi, J. Carbon nanotube-graphene hybrids for soft electronics, sensors, and actuators. *Micro Nano Syst. Lett.* **2022**, *10*, 9. [CrossRef]
24. Riyajuddin, S.; Kumar, S.; Soni, K.; Gaur, S.P.; Badhwar, D.; Ghosh, K. Study of field emission properties of pure graphene-CNT heterostructures connected via seamless interface. *Nanotechnology* **2019**, *30*, 385702. [CrossRef] [PubMed]
25. Li, X.; Tang, Y.; Song, J.; Yang, W.; Wang, M.; Zhu, C.; Zhao, W.; Zheng, J.; Lin, Y. Self-supporting activated carbon/carbon nanotube/reduced graphene oxide flexible electrode for high performance supercapacitor. *Carbon* **2018**, *129*, 236–244. [CrossRef]
26. Tang, C.; Zhang, Q.; Zhao, M.; Tian, G.; Wei, F. Resilient aligned carbon nanotube/graphene sandwiches for robust mechanical energy storage. *Nano Energy* **2014**, *7*, 161–169. [CrossRef]
27. Sheka, E.F.; Chernozatonskii, L.A. Graphene-Carbon Nanotube Composites. *J. Comp. Theor. Nanosci.* **2010**, *7*, 1814–1824. [CrossRef]
28. Slepchenkov, M.M.; Shmygin, D.S.; Zhang, G.; Glukhova, O.E. Controlling the electronic properties of 2D/3D pillared graphene and glass-like carbon via metal atom doping. *Nanoscale* **2019**, *11*, 16414–16427. [CrossRef]
29. Gong, J.; Yang, P. Investigation on field emission properties of graphene–carbon nanotube composites. *RSC Adv.* **2014**, *4*, 19622–19628. [CrossRef]
30. Matsumoto, T.; Saito, S. Geometric and Electronic Structure of New Carbon-Network Materials: Nanotube Array on Graphite Sheet. *J. Phys. Soc. Jpn.* **2002**, *71*, 2765–2770. [CrossRef]
31. Mao, Y.; Zhong, J. The computational design of junctions by carbon nanotube insertion into a graphene matrix. *New J. Phys.* **2009**, *11*, 093002. [CrossRef]
32. Novaes, F.D.; Rurali, R.; Ordejon, P. Electronic Transport between Graphene Layers Covalently Connected by Carbon Nanotubes. *ACS Nano* **2010**, *4*, 7596–7602. [CrossRef]
33. Chen, J.; Walther, J.H.; Koumoutsakos, P. Covalently Bonded Graphene-Carbon Nanotube Hybrid for High-Performance Thermal Interfaces. *Adv. Funct. Mater.* **2015**, *25*, 7539–7545. [CrossRef]
34. Varshney, V.; Patnaik, S.S.; Roy, A.K.; Froudakis, G.; Farmer, B.L. Modeling of Thermal Transport in Pillared-Graphene Architectures. *ACS Nano* **2010**, *4*, 1153–1161. [CrossRef]

35. Zhang, Z.; Kutana, A.; Roy, A.; Yakobson, B.I. Nanochimneys: Topology and Thermal Conductance of 3D Nanotube–Graphene Cone Junctions. *J. Phys. Chem. C* **2017**, *121*, 1257–1262. [CrossRef]
36. Artyukh, A.A.; Chernozatonskii, L.A.; Sorokin, P.B. Mechanical and electronic properties of carbon nanotube–graphene compounds. *Phys. Status Solidi (b)* **2010**, *247*, 2927–2930. [CrossRef]
37. Ivanovskaya, V.V.; Zobelli, A.; Wagner, P.; Heggie, M.I.; Briddon, P.R.; Rayson, M.J.; Ewels, C.P. Low-energy termination of graphene edges via the formation of narrow nanotubes. *Phys. Rev. Lett.* **2011**, *107*, 065502. [CrossRef] [PubMed]
38. Akhukov, M.A.; Yuan, S.; Fasolino, A.; Katsnelson, M.I. Electronic, magnetic and transport properties of graphene ribbons terminated by nanotubes. *New J. Phys.* **2012**, *14*, 123012. [CrossRef]
39. Cook, B.G.; French, W.R.; Varga, K. Electron transport properties of CNT–graphene contacts. *Appl. Phys. Lett.* **2012**, *101*, 153501. [CrossRef]
40. Srivastava, J.; Gaur, A. A tight-binding study of the electron transport through single-walled carbon nanotube-graphene hybrid nanostructures. *J. Chem. Phys.* **2021**, *155*, 244104. [CrossRef] [PubMed]
41. Felix, A.B.; Pacheco, M.; Orellana, P.; Latgé, A. Vertical and In-Plane Electronic Transport of Graphene Nanoribbon/Nanotube Heterostructures. *Nanomaterials* **2022**, *12*, 3475. [CrossRef]
42. Glukhova, O.E.; Nefedov, I.S.; Shalin, A.S.; Slepchenkov, M.M. New 2D graphene hybrid composites as an effective base element of optical nanodevices. *Beilstein J. Nanotechnol.* **2018**, *9*, 1321–1327. [CrossRef] [PubMed]
43. Advincula, P.A.; Beckham, J.L.; Choi, C.H.; Chen, W.; Han, Y.; Kosynkin, D.V.; Lathem, A.; Mayoral, A.; Yacaman, M.J.; Tour, J.M. Tunable Hybridized Morphologies Obtained through Flash Joule Heating of Carbon Nanotubes. *ACS Nano* **2023**, *17*, 2506–2516. [CrossRef] [PubMed]
44. Li, Y.Y.; Ai, Q.Q.; Mao, L.N.; Guo, J.X.; Gong, T.X.; Lin, Y.; Wu, G.T.; Huang, W.; Zhang, X.S. Hybrid strategy of graphene/carbon nanotube hierarchical networks for highly sensitive, flexible wearable strain sensors. *Sci. Rep.* **2021**, *11*, 21006. [CrossRef] [PubMed]
45. Shin, D.H.; You, Y.G.; Jo, S.I.; Jeong, G.H.; Campbell, E.E.B.; Chung, H.J.; Jhang, S.H. Low-Power Complementary Inverter Based on Graphene/Carbon-Nanotube and Graphene/MoS2 Barristors. *Nanomaterials* **2022**, *12*, 3820. [CrossRef] [PubMed]
46. He, Z.; Wang, K.; Yan, C.; Wan, L.; Zhou, Q.; Zhang, T.; Ye, X.; Zhang, Y.; Shi, F.; Jiang, S.; et al. Controlled Preparation and Device Application of Sub-5 nm Graphene Nanoribbons and Graphene Nanoribbon/Carbon Nanotube Intramolecular Heterostructures. *ACS Appl. Mater. Interfaces* **2023**, *15*, 7148–7156. [CrossRef]
47. McDaniel, J.G. Capacitance of Carbon Nanotube/Graphene Composite Electrodes with [BMIM+][BF4–]/Acetonitrile: Fixed Voltage Molecular Dynamics Simulations. *J. Phys. Chem. C* **2022**, *126*, 5822–5837. [CrossRef]
48. Xu, T.; Jiang, J. On the configuration of the graphene/carbon nanotube/graphene van der Waals heterostructure. *Phys. Chem. Chem. Phys.* **2023**, *25*, 5066–5072. [CrossRef]
49. Wei, L.; Zhang, L. Atomic Simulations of (8,0) CNT-Graphene by SCC-DFTB Algorithm. *Nanomaterials* **2022**, *12*, 1361. [CrossRef]
50. Zhang, S.; Kang, L.; Wang, X.; Tong, L.; Yang, L.; Wang, Z.; Qi, K.; Deng, S.; Li, Q.; Bai, X.; et al. Arrays of horizontal carbon nanotubes of controlled chirality grown using designed catalysts. *Nature* **2017**, *543*, 234–238. [CrossRef]
51. Elstner, M.; Seifert, G. Density functional tight binding. *Phil. Trans. R. Soc. A* **2014**, *372*, 20120483. [CrossRef]
52. DFTB+ Density Functional Based Tight Binding (and More). Available online: https://dftbplus.org/ (accessed on 10 June 2022).
53. Spiegelman, F.; Tarrat, N.; Cuny, J.; Dontot, L.; Posenitskiy, E.; Martí, C.; Simon, A.; Rapacioli, M. Density-functional tight-binding: Basic concepts and applications to molecules and clusters. *Adv. Phys. X* **2020**, *5*, 1710252. [CrossRef] [PubMed]
54. Mulliken, R.S. Electronic Population Analysis on LCAO–MO Molecular Wave Functions, I. *J. Chem. Phys.* **1955**, *23*, 1833. [CrossRef]
55. Marconcini, P.; Macucci, M. Transport Simulation of Graphene Devices with a Generic Potential in the Presence of an Orthogonal Magnetic Field. *Nanomaterials* **2022**, *12*, 1087. [CrossRef] [PubMed]
56. Datta, S. *Quantum Transport: Atom to Transistor*, 2nd ed.; Cambridge University Press: New York, NY, USA, 2005; pp. 217–251.
57. Glukhova, O.E.; Shmygin, D.S. The electrical conductivity of CNT/graphene composites: A new method for accelerating transmission function calculations. *Beilstein J. Nanotechnol.* **2018**, *9*, 1254–1262. [CrossRef]
58. Symalla, F.; Shallcross, S.; Beljakov, I.; Fink, K.; Wenzel, W.; Meded, V. Band-gap engineering with a twist: Formation of intercalant superlattices in twisted graphene bilayers. *Phys. Rev. B* **2015**, *91*, 205412. [CrossRef]
59. Tristán-López, F.; Morelos-Gómez, A.; Vega-Díaz, S.M.; García-Betancourt, M.L.; Perea-López, N.; Elías, A.L.; Muramatsu, H.; Cruz-Silva, R.; Tsuruoka, S.; Kim, Y.A.; et al. Large Area Films of Alternating Graphene–Carbon Nanotube Layers Processed in Water. *ACS Nano* **2013**, *7*, 10788–10798. [CrossRef]
60. Liu, H.; Deshmukh, A.; Salowitz, N.; Zhao, J.; Sobolev, K. Resistivity Signature of Graphene-Based Fiber-Reinforced Composite Subjected to Mechanical Loading. *Front. Mater.* **2022**, *9*, 818176. [CrossRef]
61. Wakabayashi, K.; Sasaki, K.I.; Nakanishi, T.; Enoki, T. Electronic states of graphene nanoribbons and analytical solutions. *Sci. Technol. Adv. Mater.* **2010**, *11*, 054504. [CrossRef]

Disclaimer/Publisher's Note: The statements, opinions and data contained in all publications are solely those of the individual author(s) and contributor(s) and not of MDPI and/or the editor(s). MDPI and/or the editor(s) disclaim responsibility for any injury to people or property resulting from any ideas, methods, instructions or products referred to in the content.

Article

High Gas Response Performance Based on Reduced Graphene Oxide/SnO₂ Nanowires Heterostructure for Triethylamine Detection

Ruiqin Peng [1], Xuzhen Zhuang [1], Yuanyuan Li [2], Zhiguo Yu [1] and Lijie Ci [2,*]

[1] School of Intelligence Engineering, Shandong Management University, Jinan 250357, China
[2] School of Materials Science and Engineering, Harbin Institute of Technology (Shenzhen), Shenzhen 518055, China
* Correspondence: cilijie@hit.edu.cn

Abstract: SnO_2 nanowires are locally synthesized by a simple thermal evaporation method and its growth mechanism is confirmed. Here, we present a simple strategy for realizing reduced graphene oxide (RGO)/SnO_2 nanowires heterostructure. As expected, the heterostructure gas-sensing response is up to 63.3 when the gas concentration of trimethylamine (TEA) is 50 ppm, and it exhibits an excellent dynamic response with high stability at 180 °C. A low detection limit of 50 ppb level is fully realized. Compared to SnO_2 nanowires, the sensing performance of the RGO/SnO_2 heterostructure-based sensor is greatly enhanced, which can be ascribed to the RGO and the heterostructure. The RGO/SnO_2 composite engineering poses an easy way to make full use of the advantages originating from RGO and heterostructure.

Keywords: RGO/SnO_2; heterostructure; gas response; mechanism

Citation: Peng, R.; Zhuang, X.; Li, Y.; Yu, Z.; Ci, L. High Gas Response Performance Based on Reduced Graphene Oxide/SnO₂ Nanowires Heterostructure for Triethylamine Detection. *Coatings* 2023, *13*, 849. https://doi.org/10.3390/coatings13050849

Academic Editor: Keith J. Stine

Received: 28 March 2023
Revised: 24 April 2023
Accepted: 27 April 2023
Published: 29 April 2023

Copyright: © 2023 by the authors. Licensee MDPI, Basel, Switzerland. This article is an open access article distributed under the terms and conditions of the Creative Commons Attribution (CC BY) license (https://creativecommons.org/licenses/by/4.0/).

1. Introduction

Owing to the great demands of the chemical industry and real-time gas monitoring systems, high sensitivity and good stability of gas sensors are attracting tremendous attention. SnO_2, ZnO, and WO_3 have always been the traditional and dominant materials for sensor fabrication [1–6]. However, the excellent stability, long-cycles, and low detection limit of the sensors still face the challenge, especially working at a high temperature above 200 °C. At higher temperatures, the inevitable grain growth of materials would degrade the sensor stability and life [6]. Researchers have tried to implement some strategies (e.g., low dimensional nanostructures, metal doping, heterostructure engineering, etc.) for enhancing gas-sensing properties [7–9]. One-dimensional SnO_2 nanowire has a large surface-to-volume ratio and constant carrier screening length, which makes them more sensitive and efficient than SnO_2 film to transduce surface chemical processes into electrical signals [10]. Zou et al. demonstrated a hollow SnO_2 microfiber as a sensing layer and found a high response to TEA gas. However, the optimal operating temperature is up to 270 °C [11].

As a promising sensing material and derivative material of the graphene family, the RGO layer can provide more adopted active sites and fast gas diffusion. It normally acts as a p-type semiconductor, exhibiting the unique advantages of gas sensitive at room temperature [12]. Therefore, it has exhibited great potential for enhancing sensor performance by combing the merits of the one-dimensional SnO_2 nanowire and RGO material. For example, Song et al., reported a sensitive SnO_2/RGO nanocomposite H_2S gas sensor and the optimal sensor response was 33 in 2 s at 50 ppm of H_2S, which showed great potential in application [13]. Zhang et al. presented the synthesis of Ag/SnO_2/RGO ternary nanocomposites, and the composite sensors exhibited high response to TEA gas at 220 °C [14]. However,

RGO/SnO$_2$ crystalline nanowires composite gas sensors for TEA detection have been reported very little.

Here, we reported on the RGO/SnO$_2$ nanowires composite heterostructure for TEA gas sensing, which fully integrated the advantages of the RGO and SnO$_2$ nanowire by forming RGO/SnO$_2$ heterostructure and enhanced the gas-sensing properties. When the concentration of TEA gas was 50 ppm, the response was up to 63.3 and it exhibited good stability. The sensors were more sensitive to TEA gas than other gases. The sensing performance of the RGO/SnO$_2$ nanowires composite heterostructure is enhanced compared to that of SnO$_2$ nanowires; the mechanism is also discussed.

2. Materials and Methods

2.1. Preparation of RGO/SnO$_2$ Nanowires Composite Heterostructure

Synthesis of RGO. Graphene oxide (GO) water dispersion was purchased from Suzhou Tanfeng Graphene Tech. Inc. (Suzhou, China) The GO (1 mg/25 mL) was reduced into RGO by a chemically reducing method [15]. Briefly, the Hydroxylamine Hydrochloride (0.2 g) was added into the GO dispersion. The mixture was stirred for 6 h and then was under ultrasonic treatment for 1 h. After that, the dispersions were transferred into a 50 mL Teflon-lined stainless steel autoclave and maintained at 100 °C for 12 h. The black product was filtered and rinsed by acetone, ethanol, and deionized water to obtain the pure RGO powder.

Fabrication of SnO$_2$ nanowires and sensors. The SnO$_2$ nanowires were prepared by thermal evaporation method. The SnO$_2$ nanowires are fabricated using the above procedures as shown in Figure 1a, which shows the growth conditions of SnO$_2$ nanowires. Firstly, Sn powder was placed on the quartz plate, and SiO$_2$/Si substrates (1 × 1 cm^2) with Au film (5 nm) were placed about 5 cm away from the Sn powder. Then, we placed the quartz piece in the center of the tube furnace. Ar gas (20 standard cubic centimeter per minute (sccm)) was introduced into the tube and the furnace pressure was constant at about 100 Pa. The furnace temperature was raised to 1000 °C by 10 °C/min. The nanowires growth period was about 20 min. The Au film pattern staying on the substrate was confined by the standard lithography process. The interdigitated Au-film electrode of the sensors was a planar device structure and the thickness was about 200 nm. Then, the RGO powder was fully dispersed in N, N-dimethylformamide (DMF) (0.5 mg/mL) by ultrasonication treatment for 2 h and was coated onto the surface of SnO$_2$ nanowires by spinning coating. Afterward, the samples were dried in vacuum oven at 60 °C for 4 h and then annealed at 200 °C for 60 min under Ar atmosphere to remove residual DMF. Finally, gas sensors were aged at 120 °C for 24 h. Figure 1b details the fabricating processes of RGO/SnO$_2$ composite-based sensor.

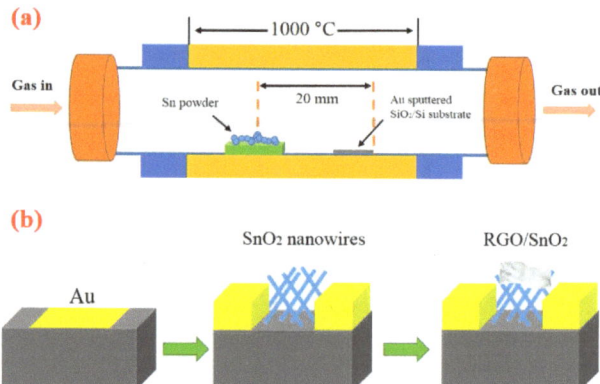

Figure 1. (**a**) Schematic of the synthesizing process of the SnO$_2$ nanowires; (**b**) the RGO/SnO$_2$ composite heterostructure-based sensor preparation process.

2.2. Materials Characterization and Sensor Properties

The microstructure and morphology were characterized by the FEI QUANTA 200 scanning electron microscope (SEM, FEI, Hillsboro, OR, America). The transmission electron microscope (TEM, JEM-2100, JEOL, Tokyo, Japan) with the selected area electron diffraction (SAED) and energy-dispersive X-ray spectroscopy (EDS). The crystalline structure was identified by MiniFlex 600 X-ray powder diffraction (XRD, Rigaku, Tokyo, Japan) with Cu Kα1 radiation (λ = 1.54056 Å). The surface chemistries were identified by Raman spectrometer (LabRAM HR800, Horbiba Jobin Yvon, Paris, France). The gas-sensing properties were tested using a commercial CGS-4TP analysis system. The test system has a closed chamber with a capacity of 18 L. The TEA solution was injected and was heated as gas molecular transport to the surface of the sensor. The gas response was defined as using the ratio of the sensor resistance in air (R_a) to that in the target gas (R_g). The sensing properties were tested at temperatures from 25 to 250 °C. The response and recovery time were calculated as 90% of the maximum after injecting target gas and backing to air [16]. The relative humidity was kept constant at 50%~60%.

3. Results and Discussion

Figure 2 shows the morphology and microstructure of the RGO/SnO$_2$ composite. In Figure 2a,b, the RGO layer adhered onto the surface of SnO$_2$ nanowires. The nanowires are long and straight. The diameter of the nanowire is around 50 nm and the length is up to 50 µm. The existing network structure of SnO$_2$ nanowires gets a large surface-to-volume ratio and numerous nanowire–nanowire junctions. Figure 2c–h shows the TEM image, SAED pattern, and EDS spectrum of the synthesized SnO$_2$ nanowires. A lattice spacing of 0.26 nm well matches to the d-spacing (101) of SnO$_2$ nanowires. The SAED and EDS data also confirmed that SnO$_2$ nanowires are successfully synthesized. As shown in Figure 2f–h, Au, Sn, and O are distributed uniformly.

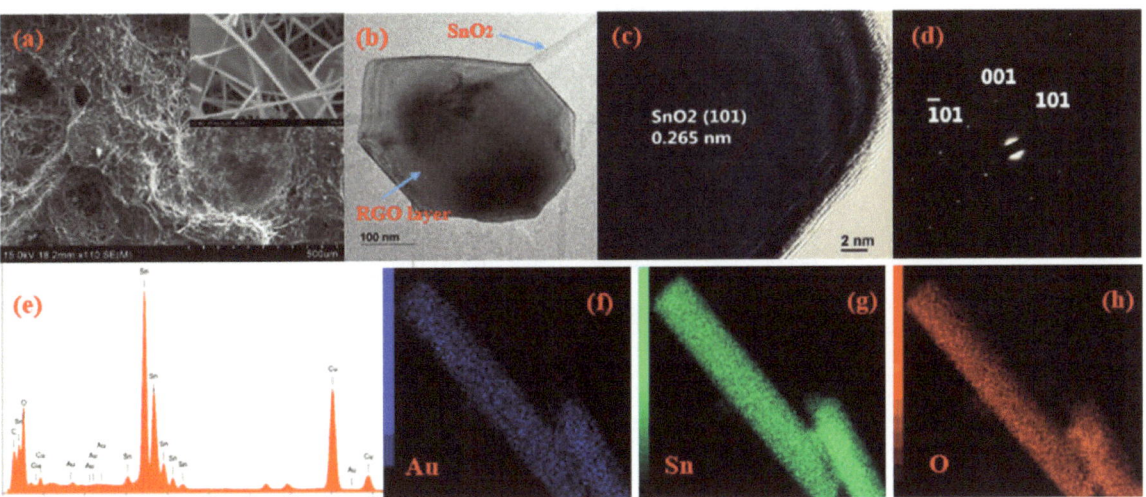

Figure 2. (a) SEM image of the RGO/SnO$_2$ composite. Inset: the enlargement of a part of the composite; (b) the TEM image of the composite; (c) the HRTEM image of a single-crystalline SnO$_2$ nanowire; (d) the SAED pattern of the SnO$_2$ nanowire; (e) the EDS spectra of the SnO$_2$ nanowire; (f–h) the EDS maps of Au, O, and Sn element of the SnO$_2$ nanowire.

As mentioned before, due to its large surface-to-volume ratio and excellent electron transport properties, the SnO$_2$ nanowire has been widely used for toxic gas detection. Commonly, the SnO$_2$ nanowire growth mechanism corresponds to vapor–liquid–solid (VLS) or vapor–solid (VS) mechanism [17,18]. VLS mechanism has a significant feature that the metal catalyst would stay at the tip of the nanowire. In this work, EDS mapping indicated Au plays a catalyst and active sites role during the preparation of SnO_2 nanowires. It revealed that the SnO_2 nanowires are mainly composed of O, Sn, and Au. We performed an easy way to show the root of the nanowire to obtain the relationship between Au and nanowires. As shown in Figure 3a, the SnO_2 nanowires were rooted in the Ni foam. The growth parameters are consistent with the SnO_2 nanowires prepared on SiO_2/Si substrate, excluding the changed substrate. We can easily recognize the responding active site for nanowires. It is conducive to speculate about the growth mechanism, which is shown in Figure 3b. Based on the above SnO_2 morphology and structure property analysis, a simple model for the growth mechanism of nanowires is discussed [19].

$$2Sn\ (L) + O_2\ (V) \rightarrow 2SnO\ (V) \tag{1}$$

$$2SnO\ (V) + Au\ \text{nano-sized droplets}\ (L) \rightarrow SnO_2\text{-}Au\ (S) + Sn \tag{2}$$

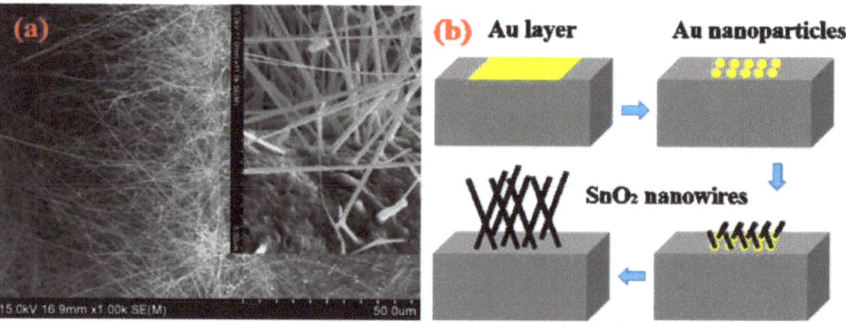

Figure 3. (a) The SEM image of the synthesized SnO_2 nanowires on Ni foam, the insetting picture shows the root of the nanowires; (b) the SnO_2 nanowires growth mechanism on SiO_2/Si substrate.

Firstly, because of the low melting point of tin (231.9 °C), the Sn powder would present in the liquid metallic [20] and the liquid tin could react with residual oxygen to yield SnO vapor. Secondly, the Au layer acted as nucleation sites on the SiO_2/Si substrate, while it provided the favored sites for absorption of SnO. Then, the SnO was transformed onto the surface of the Au liquid droplets by the carrying gas. Finally, due to the Au catalyst and SnO decomposition, the SnO_2 nanowires formed immediately [21,22]. Interestingly, there are no metallic Au at the tips of the SnO_2 nanowires but distributed in the whole nanowire as confirmed by the EDS mapping, which is different from the typical characteristics of the VLS growth mechanism.

Figure 4a shows XRD patterns of the GO, RGO, SnO_2 nanowires, and RGO/SnO_2 composite. GO shows a character (002) diffraction peak at 2θ = 10.83°. The RGO diffraction peak position is shifted to 23.85°. Hereby, the interlayer distance changed from 0.81 nm to 0.4 nm. It resulted in the decrease in the RGOs' interlayer spacing, which means that oxygen-related chemical groups were significantly removed from GO. The SnO_2 nanowires and RGO/SnO_2 composite matched well with the tetragonal rutile structure (SnO_2, JCPDS No. 41-1445). For RGO/SnO_2 composite, the RGO diffraction peak disappeared, which might be ascribed to the low RGO amount. The significant structural changes occurring from GO to the RGO are also presented in the Raman spectra. In Figure 4b, the Raman spectrum of the GO and RGO samples shows two obvious peaks at 1353 cm^{-1} (D-band) and 1595 cm^{-1} (G-band). Commonly, the G band peak can be ascribed to the E_{2g} vibrational mode of the sp2-bonded carbon atoms [23–25], while the D band peak is ascribed to the defect-induced mode [26,27]. The intensity ratio of the D to G band value of the GO is about 1.049, the increased value of RGO (1.156) suggests a successful chemical reduction of GO. For SnO_2 nanowires and a RGO/SnO_2 composite, the induction of the RGO would not change the character peaks of the SnO_2 nanowires. The three Raman scattering peaks at 474.8, 633.3, and 773.5 cm^{-1} correspond to the E_g, A_{1g}, and B_{2g} vibration modes of SnO_2 [28]. Otherwise, the peak at 498 cm^{-1} is reported to be A_{2u} mode, which has not been detected in bulk SnO_2 [18]. These character peaks further confirm the existence of the SnO_2 and RGO in the RGO/SnO_2 composite, which agrees with the results of XRD characterization.

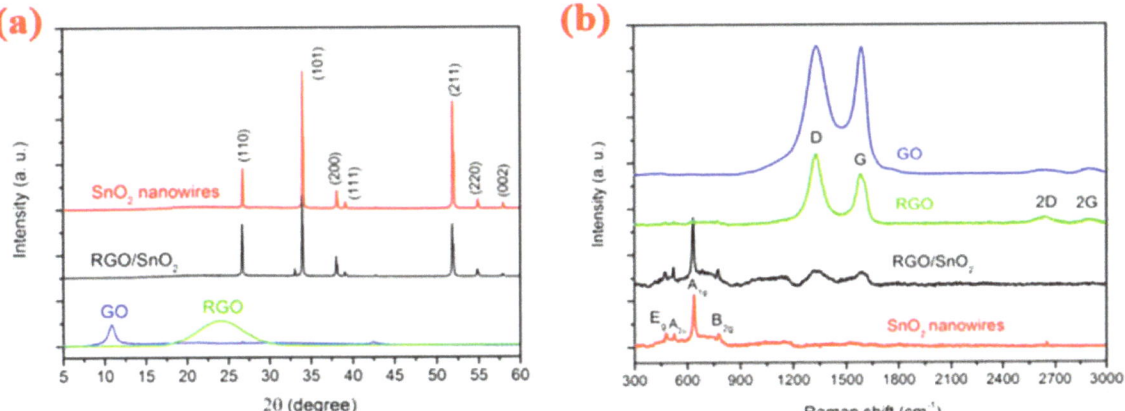

Figure 4. (**a**) The XRD results; (**b**) Raman spectra of GO, RGO, RGO/SnO_2 composite, and SnO_2 nanowires.

The sensing properties of SnO_2 nanowires and RGO/SnO_2 composite sensor toward 50 ppm TEA gas at different temperatures are shown in Figure 5a. For the SnO_2 nanowires sensor, the optimal temperature is about 200 °C. Thus, the best gas sensing temperature of the RGO/SnO_2 composite sensor is about 180 °C. RGO/SnO_2 composite sensor had the highest gas response at low temperatures due to the introduction of RGO. As expected, the gas response of the RGO/SnO_2 composite is up to 63.3, which is higher than that of SnO_2 nanowires sensor at the same testing procedures (32 under 50 ppm of TEA gas). Afterward, as the temperature increasing further, the response gradually decreases, which can be attributed to the drop down of equilibrium adsorption capacity [6]. Figure 5b presents dynamic response-recovery properties of the sensor for different TEA gas concentrations ranging from 50 ppb to 200 ppm at 180 °C. The response of the sensor is 2.6 at a TEA concentration of 100 ppb. The sensor got a low detection limit of 50 ppb level. At 180 °C, the sensor response gradually increased with the increase in the gas concentration. When towards 200 ppm, the response is up to 100, showing the excellent gas-sensing property. Figure 5c shows the response time of the sensor is about 10 s. The stability of the RGO/SnO_2 composite sensor is also measured, and the six-cycling dynamic response is shown in Figure 5d. Accordingly, the sensor shows an almost constant response time and response curves for detecting the TEA gas concentration of 1 ppm, showing good stability of the sensor.

The long-term stability and selectivity of gas sensing are fundamental factors in gas detection applications. Figure 6a shows the gas response of the RGO/SnO_2 composite based sensor to 15 dynamic cycles after 30 days at a concentration of 50 ppm. As we can see, the sensor shows good long-term stability and cycling performance. Compared to the fresh sensor, the late sensors' response value had a litter change. As shown in Figure 6b, the sensor responses to different detected gases in a continuous process. The data shows that the response intensity to 1 ppm TEA gas is almost twice as high than the response to the other six gases, even responding to a gas concentration of 100 ppm. In particular, the sensor has a weak response to water, and it shows an anti-water property, which is good for detecting TEA gas regardless of humidity in real situations.

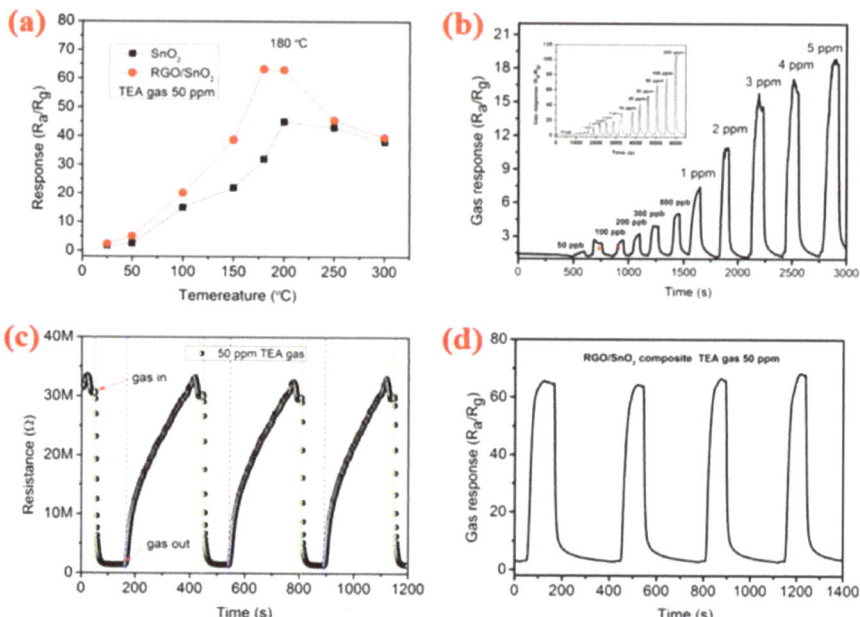

Figure 5. (**a**) Sensor response of the SnO$_2$ nanowires and RGO/SnO$_2$ composite to 50 ppm TEA gas at different working temperatures; (**b**) the dynamic gas response at different TEA gas concentrations ranging from 50 ppb to 200 ppm; (**c**) sensor response time toward 50 ppm TEA gas at 180 °C; (**d**) the four dynamic gas-sensing response cycles at 180 °C.

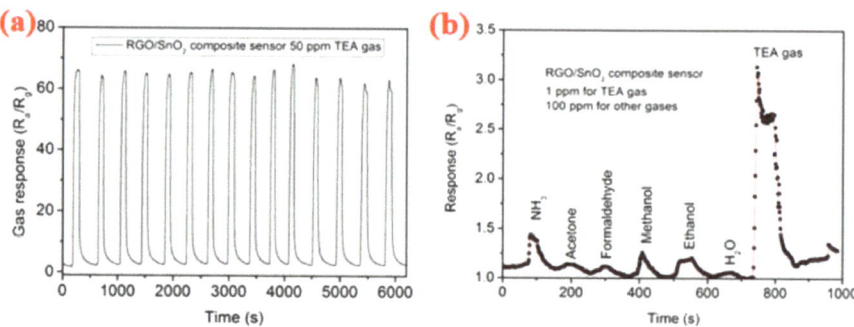

Figure 6. (**a**) The gas sensing stability of the sensor after 30 days; (**b**) the gas selectivity of the sensor to six different gases at a concentration of 1 ppm.

Generally, it is well known that the one-dimensional SnO$_2$ nanowire has a large surface-to-volume ratio and constant carrier screening length, which facilitates more rapid and high-efficient adsorption for improving the gas response [29]. Thus, due to the intertwined and irregular stacking of SnO$_2$ nanowires, the material resistance is as high as hundreds of megaohms (MΩ), making it difficult to integrate with existing electronic systems. The RGO could help improve the conductivity of metal oxide and promote the transfer of electrons. When the composite material meets air, oxygen would be adsorbed, and then would react with the free electrons to ionize oxygen into oxygen species (O^{-2}, O^{2-}, and O^-) on the surface of SnO$_2$ nanowires and RGO [30]. The depletion layer formed as shown in Figure 7a. As exposure to targeted gas, the captured electrons are released as reacting with the existing oxygen ions, reducing the thickness of the depletion layer and the resistance would be minimized. The gas response is largely enhanced. Wang et al. demonstrated the preparation of ZnO/SnO$_2$ heterostructure on RGO layer, and found promising sensing behavior upon

NO₂ exposure even at room temperature. They presented that the heterostructure could offer effective electronic interaction and high transfer efficiency of charges at the interface, which help to adsorb oxygen for enhancing the gas response [31]. So, in Figure 7b, we showed the current–voltage characteristic relationship of the RGO/SnO₂ composite gas sensor in air and after exposure to 50 ppm TEA gas, respectively. After being exposed to targeted gas, the current proudly increased because of the changed conductivity after surface reactions between TEA gas and adsorption oxygen. The following reaction can be depicted as follows [32]:

$$(C_2H_5)_3N + xO^{\delta-} \rightarrow H_2O + CO_2 + NO_2 + xe^{-1} \tag{3}$$

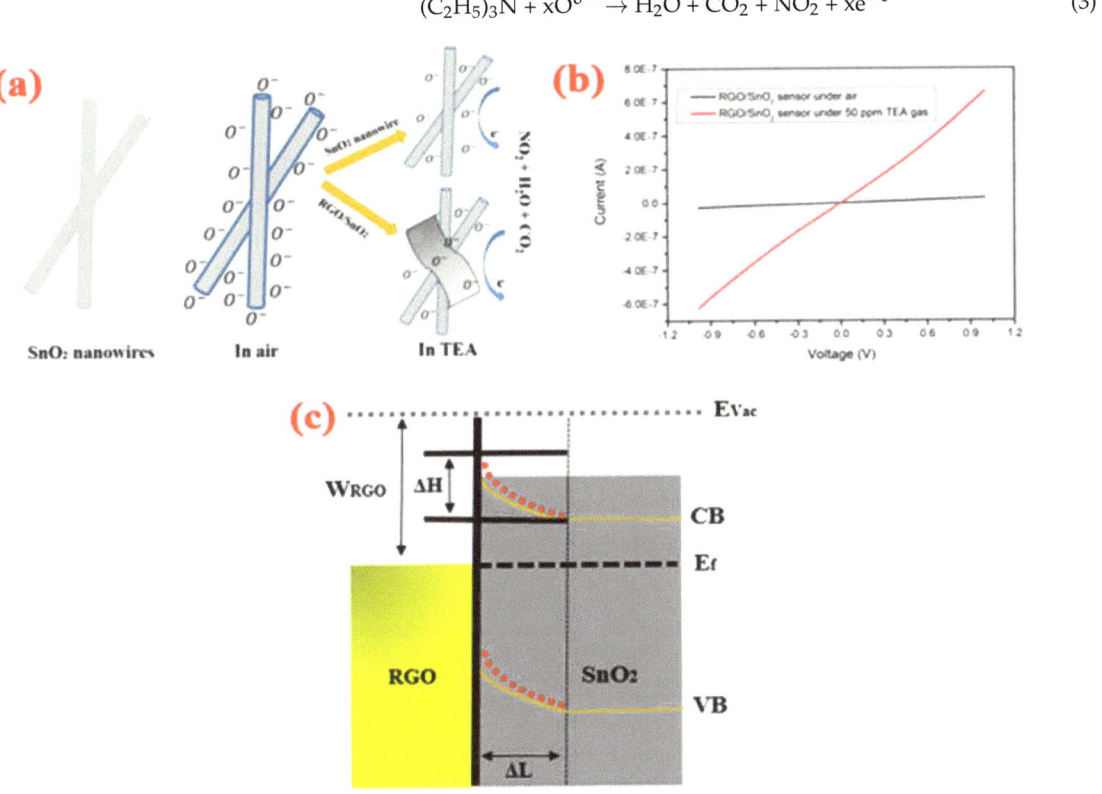

Figure 7. (**a**) Schematic of the RGO/SnO₂ composite gas-sensing mechanism; (**b**) the current–voltage characteristic relationship of the RGO/SnO₂ composite gas sensor in air and after exposure to 50 ppm TEA gas, respectively; (**c**) energy band diagram of the RGO/SnO₂ composite.

The gas-sensing mechanism of RGO/SnO₂ heterostructure-based sensors can be concluded. Notably, RGO could play a significant role and provide reactive centers for adsorption, which facilitates rapid electrons transformation between targeted gas and SnO₂. SnO₂, as a promising and mature gas-sensing material, could also promote effective charge transfer and oxygen adsorption. Therefore, when the RGO/SnO₂ composite encounters TEA gas, the energy band bending would happen until balance, as depicted in Figure 7c. The electron movement will start flowing from RGO to SnO₂. So, when encountering the targeted gas, the adsorbed process was happening at the interface, a small change in the potential barrier (E) would contribute to a large resistance change of the material. The resistance is related to the height of the potential barrier, which is followed by the equation [30]:

$$R = R_0 \exp(qE/kT) \tag{4}$$

Generally, R_0 is the initial resistance, q is the electron charge, E is the height of the potential energy barrier, k is Boltzmann's constant, T is the temperature, and R is the overall resistance of the material. So, based on the above results, the RGO/SnO$_2$ heterostructure enhanced the gas-sensing performance and provided a reliable strategy to obtain the highly sensitive.

4. Conclusions

In summary, the crystalline SnO$_2$ nanowires are synthesized by a simple thermal evaporation method. We also presented a coating technology to realize RGO/SnO$_2$ heterostructure. When the TEA gas concentration is 50 ppm, the gas response is raised to 63.3, and it showed an excellent dynamic response with a high stability at 180 °C. A lower detection limit of 50 ppb level is achieved. Compared to SnO$_2$ nanowires, the sensing performance of the RGO/SnO$_2$ heterostructure-based sensor is greatly enhanced, which can be ascribed to the RGO and the heterostructure. This heterostructure engineering poses an easy and repeatable way for achieving highly sensitive performance.

Author Contributions: Formal analysis, investigation, writing—original draft preparation. R.P.; writing—review and editing, X.Z. and Y.L.; visualization, Z.Y.; supervision, L.C. All authors have read and agreed to the published version of the manuscript.

Funding: This work was supported by the Shandong Provincial Science and Technology Major Project (2019GGX101029) and the National Science Foundation of Shandong Province (ZR2017BEM049).

Institutional Review Board Statement: Not applicable.

Informed Consent Statement: Not applicable.

Data Availability Statement: Not applicable.

Acknowledgments: The authors thank the technical supporting from the Research Center for Carbon Nanomaterials, Shandong University.

Conflicts of Interest: The authors declare no conflict of interest.

References

1. Wang, D.; Chu, X.; Gong, M. Gas-Sensing Properties of Sensors Based on Single-Crystalline SnO$_2$ Nanorods Prepared by a Simple Molten-Salt Method. *Sens. Actuators B Chem.* **2006**, *117*, 183–187. [CrossRef]
2. Shi, L.; Naik, A.J.; Goodall, J.B.; Tighe, C.; Gruar, R.; Binions, R.; Parkin, I.; Darr, J. Highly Sensitive ZnO Nanorod-and Nanoprism-Based NO$_2$ Gas Sensors: Size and Shape Control Using a Continuous Hydrothermal Pilot Plant. *Langmuir* **2013**, *29*, 10603–10609. [CrossRef]
3. Tomer, V.K.; Devi, S.; Malik, R.; Nehra, S.; Duhan, S. Highly Sensitive and Selective Volatile Organic Amine (VOA) Sensors Using Mesoporous WO$_3$–SnO$_2$ Nanohybrids. *Sens. Actuators B Chem.* **2016**, *229*, 321–330. [CrossRef]
4. Barthod-Malat, B.; Hauguel, M.; Behlouli, K.; Grisel, M.; Savary, G. Influence of the Compression Molding Temperature on VOCs and Odors Produced from Natural Fiber Composite Materials. *Coatings* **2023**, *13*, 371. [CrossRef]
5. Lazau, C.; Nicolaescu, M.; Orha, C.; Pop, A.; Căprărescu, S.; Bandas, C. In Situ Deposition of Reduced Graphene Oxide on Ti Foil by a Facile, Microwave-Assisted Hydrothermal Method. *Coatings* **2022**, *12*, 1805. [CrossRef]
6. Zhu, L.Y.; Miao, X.Y.; Ou, L.X.; Mao, L.W.; Yuan, K.; Sun, S.; Devi, A.; Lu, H.L. Heterostructured α-Fe$_2$O$_3$@ ZnO@ ZIF-8 Core–Shell Nanowires for a Highly Selective MEMS-Based ppb-Level H$_2$S Gas Sensor System. *Small* **2022**, *18*, 2204828. [CrossRef] [PubMed]
7. Kim, T.; Cho, W.; Kim, B.; Yeom, J.; Kwon, Y.M.; Baik, J.M.; Kim, J.J.; Shin, H. Batch Nanofabrication of Suspended Single 1D Nanoheaters for Ultralow-Power Metal Oxide Semiconductor-Based Gas Sensors. *Small* **2022**, *18*, 2204078. [CrossRef]
8. Yu, Y.-T.; Dutta, P. Examination of Au/SnO$_2$ Core-Shell Architecture Nanoparticle for Low Temperature Gas Sensing Applications. *Sens. Actuators B Chem.* **2011**, *157*, 444–449. [CrossRef]
9. Yu, Q.; Zhu, J.; Xu, Z.; Huang, X. Facile Synthesis of α-Fe$_2$O$_3$@ SnO$_2$ Core–Shell Heterostructure Nanotubes for High Performance Gas Sensors. *Sens. Actuators B Chem.* **2015**, *213*, 27–34. [CrossRef]
10. Wang, B.; Zhu, L.; Yang, Y.; Xu, N.; Yang, G. Fabrication of a SnO$_2$ Nanowire Gas Sensor and Sensor Performance for Hydrogen. *J. Phys. Chem. C* **2008**, *112*, 6643–6647. [CrossRef]
11. Zou, Y.; Chen, S.; Sun, J.; Liu, J.; Che, Y.; Liu, X.; Zhang, J.; Yang, D. Highly Efficient Gas Sensor Using a Hollow SnO$_2$ Microfiber for Triethylamine Detection. *ACS Sens.* **2017**, *2*, 897–902. [CrossRef] [PubMed]
12. Lipatov, A.; Varezhnikov, A.; Wilson, P.; Sysoev, V.; Kolmakov, A.; Sinitskii, A. Highly Selective Gas Sensor Arrays Based on Thermally Reduced Graphene Oxide. *Nanoscale* **2013**, *5*, 5426–5434. [CrossRef] [PubMed]

cell consists of six sulfur atoms and three molybdenum atoms in a trigonal crystal system. The lattice constants are a = b = 3.190 Å, and the space group is R$\bar{3}$m [47]. From the crystal structure diagram, the structure of this material can be considered as a network of herringbone Mo-S atomic chains in two directions, one along the monoclinic lattice and the other perpendicular to the lattice direction. The angles of S-Mo-S along these two directions are 81.567° and 98.433°, respectively (See Figure 1e).

To investigate how the atomic geometry affects PR, three additional materials were selected for comparison: the MoS$_2$ with zigzag structure (space group P6$_3$/mmc), the MoS$_2$ without the special undulations (space group F$\bar{4}$3m), and the WS$_2$ with the same R$\bar{3}$m space group but different elements. The calculated results are plotted in Figures 2 and S2. The lattice constants and the atomic coordinates in every cell after optimization are included in Table S2. Clearly, among the three compared materials, MoS$_2$ with space group F$\bar{4}$3m has no obvious zigzag geometry, and it shows a positive Poisson's ratio (PPR). This is in agreement with the previous conclusion that the unique sawtooth-like geometry is necessary for an NPR to occur [48]. However, the PR of the other two materials with similar sawtooth geometries (MoS$_2$, space group P6$_3$/mmc; WS$_2$, space group R$\bar{3}$m) is also positive. This suggests the NPR is not only related to the material geometry. Other factors may play a role too.

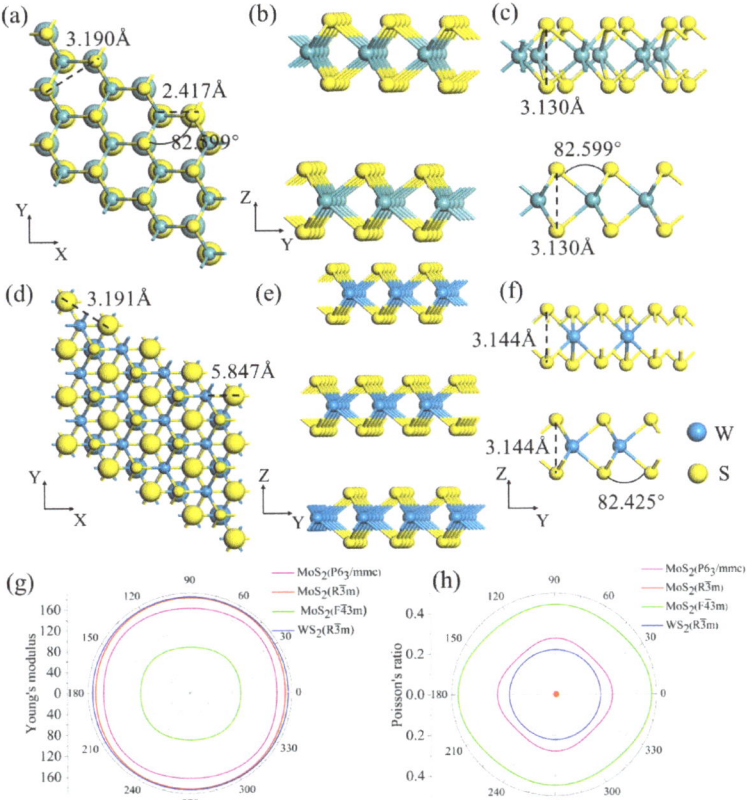

Figure 2. (**a**) Structural snapshots (top view) of MoS$_2$ (space group P6$_3$/mmc). (**b**) Side view of MoS$_2$ (space group P6$_3$/mmc). (**c**) Local enlargement of a side view of the monolayer MoS$_2$ (space group P6$_3$/mmc). (**d**) Structural snapshots (top view) of WS$_2$ (space group R$\bar{3}$m). (**e**) Side view of WS$_2$ (space group R$\bar{3}$m). (**f**) Local enlargement of the side view of the monolayer WS$_2$ (space group R$\bar{3}$m). (**g**) Summary graph of Young's modulus and (**h**) PR of four materials. To show their structure more clearly, the above materials are enlarged for the part of the atoms that are obscured between the layers; this does not mean that they are different in size.

To gain a more in-depth insight into the deformation mechanism behind the NPR, we further analyzed the electronic interactions. We calculated the PDOS of the four materials to observe their electronic structures, as shown in Figure 3. The PDOS peak shape and position of the 4d orbital of the Mo atom and the 3p orbit of the S atom are similar in energy, indicating the strong orbital interaction of the Mo atom and S atom (the yellow highlighted portion of Figure 3a). It can be seen that both sets of orbitals overlap above the Fermi energy level, indicating that the atoms are antibonding. It is well-known that the bonding state leads to attractive interactions, while the possession of the antibonding state causes repulsive interactions. This is most likely a key factor in the generation of NPR for MoS_2. Moreover, the MoS_2 that shows an NPR has an overlap area of 5.854 above the Fermi level, while the other three materials showing PPR have overlaps of 1.807, 0, and 2.508 above the Fermi level, respectively, which all indicate that their coupling is not as strong as that of the NPR materials at the p-d orbitals.

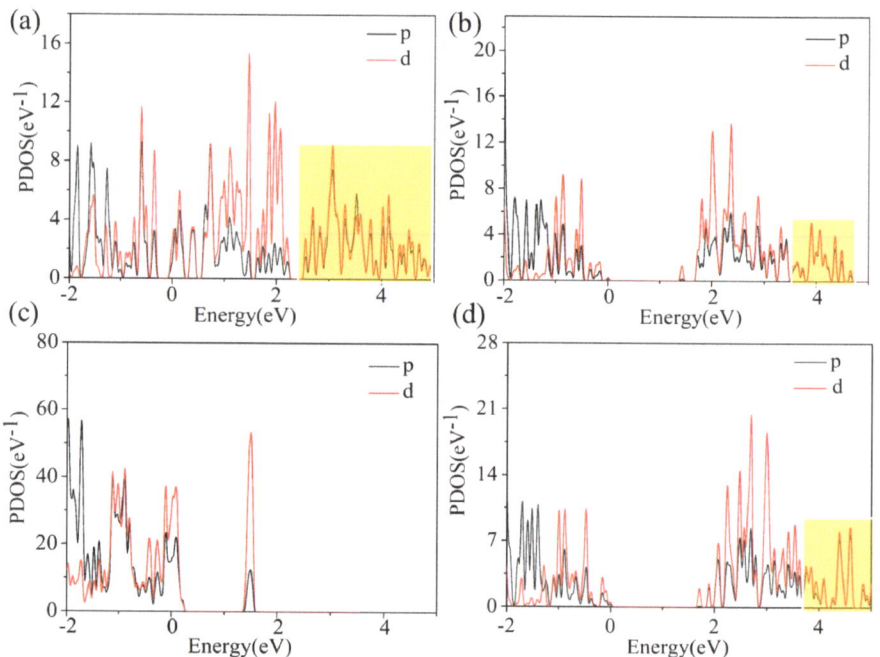

Figure 3. (**a**) PDOS of MoS_2(space group $R\bar{3}m$). The p-d orbital coupling is reflected in the overlap of its PDOS. The area of overlap above the Fermi level is 5.854. (**b**) PDOS of MoS_2 (space group $P6_3/mmc$). The area of overlap above the Fermi level is 1.807. (**c**) PDOS of MoS_2(space group $F\bar{4}3m$). The area of overlap above the Fermi level is 0. (**d**) PDOS of WS_2(space group $R\bar{3}m$). The area of overlap above the Fermi level is 2.508. The Fermi level is set to 0. The yellow highlighted part indicates the overlapping part of the PDOS of the materials.

Figure 3d indicates that the interaction of the p orbital and d orbital of WS_2 is significantly weaker than MoS_2 in the energy range of 0 to 6 eV. Although both MoS_2 (space group $R\bar{3}m$) and WS_2 (space group $R\bar{3}m$, and the value of PR is 0.25 in Figure 2h) share the same space group, there are differences in their electronic structures that lead to very different results. Therefore, the electronic structure is one of the characteristics of the NPR of MoS_2. In summary, an NPR material should have a special geometry in terms of structure (i.e., undulating structure), while in microscopic terms, it should have a strong enough coupling between different orbitals (i.e., electron interaction), and the atoms are anti-bonding (the atoms will tend to interact with each other in a repulsive manner). Materials that have all these characteristics will have a much higher probability of NPR.

Electronic interactions can be divided into interlayer and intralayer based on the geometric configuration of MoS$_2$. Furthermore, we investigated how the electronic structure affects the interatomic forces that lead to the NPR in the material. Figure 4a displays the structure of the initial MoS$_2$, and Figure 4b displays the structure with the new alignment rebuilt by cleaving the (0 0 1) surface of MoS$_2$. The difference between them is that the initial structure is staggered and aligned, while the modeled structure is overlapping and aligned. By varying their layer spacing, it is possible to investigate how much the interlayer forces of these two very different alignments contribute to the generation of NPR. As illustrated in Figure 4c, the PR of the material varies in value but remains negative overall. The PR varies from −0.67 to −0.82 and from −0.10 to −0.13, with a moderate variation. It indicates that the value of the PR does not change greatly with an increase in the layer spacing, indicating that the layer spacing has almost no impact on the NPR of the material. Therefore, the interatomic forces caused by the intralayer electronic structure of monolayer MoS$_2$ may be the dominant factor.

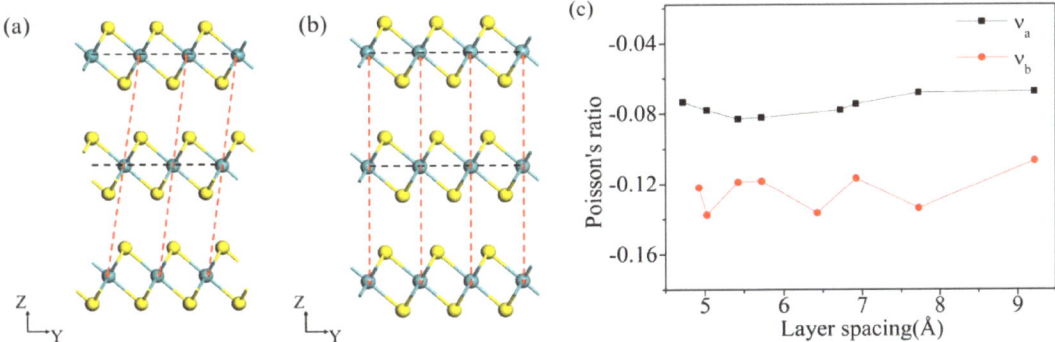

Figure 4. (**a**) The original structure of MoS$_2$. (**b**) The structure with overlapping alignment between layers of MoS$_2$. (**c**) The relationship between PR and layer spacing for two different arrangements of structures.

Based on the above analyses, the force generated by the electronic structure between the layers has little effect on the deformation of the whole structure. In the next step, we applied a strain to the single-layer MoS$_2$ (space group R$\bar{3}$m) and observed how the interaction between the electrons in the structure within the layer affected the material's PR. Figure 5a shows a sketch of the original structure with a monolayer of MoS$_2$. This is a material with a folded structure with upper and lower pyramidal shapes. Figure 5b shows a schematic of the structure after it has been stretched. When two Mo atoms, Mo$_A$ and Mo$_B$, are pulled in the Y-axis, the structure produces a force of compression in the Z-axis because of the intense interaction of the p-d orbitals, resulting in the structure being compressed in that direction and pulled in the X-axis. The bond angles of ∠CAB and ∠CBA are reduced behind the deformation. In principle, the change of ∠CAB is not impacted by the stretching of the Mo$_A$ and Mo$_B$ atoms, which is only related to the reciprocal action of the p-d orbitals. That is, the displacements occurring in the Mo$_A$ and Mo$_B$ during the deformation are only relevant to the strength inside the cell. Therefore, the angular variation of ∠CAB could be used to describe the mechanical behavior generated within the cell. When a 5% stretch strain is applied in the Y-axis, the bond angle of ∠CAB decreases from 49.216° to 45.376° (a decrease of 3.84°), and the interplanar spacings d$_{AC}$ and d$_{CE}$ decrease by 0.052 Å and 0.039 Å, respectively. The strong coupling of the p-d orbitals causes a large amount of strain energy generated during deformation to be stored in the reduced ∠CAB and ∠CBA, and this strain energy is released by increasing the distance between Mo atoms D and Mo atoms A, leading to an NPR. This confirms the above inference that the p-d orbital interactions lead to a planar NPR in monolayered MoS$_2$. Further observing the PDOS of MoS$_2$ in Figure 3a, the p-orbital overlaps with the d-orbital at positions above the Fermi level, which indicates that the metal atoms exhibit a strong antibonding property, which leads to mutual repulsion between the atoms. As a result, the NPR phenomenon appears.

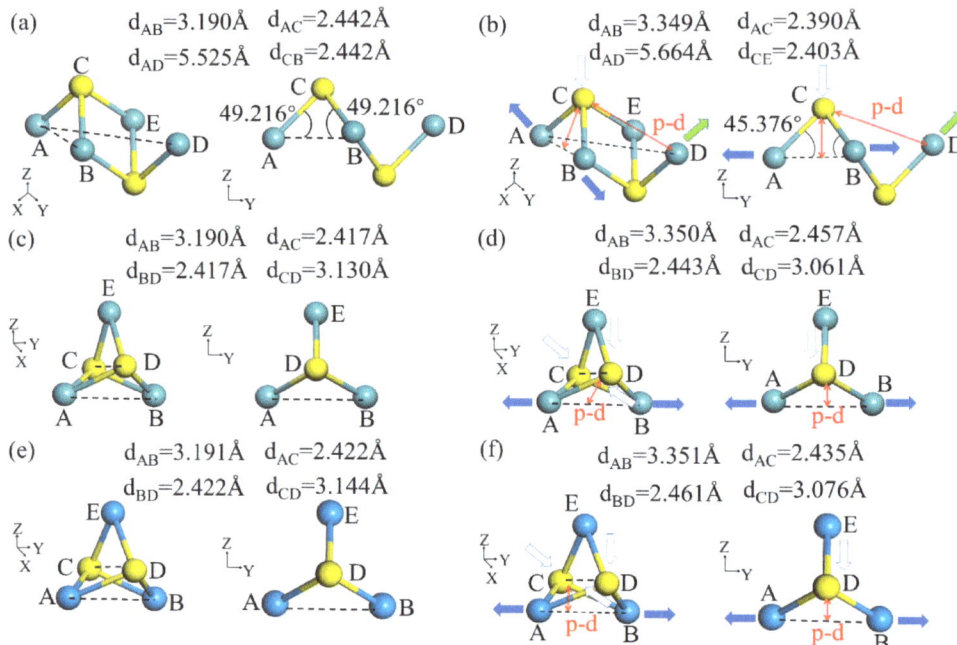

Figure 5. (**a**) Structure of single-layer MoS$_2$ (space group R$\bar{3}$m) without strain. (**b**) Structure of single-layer MoS$_2$ (space group R$\bar{3}$m) after strain is applied. The force is applied along the Y-axis. The red arrow indicates p-d orbital interaction and the green arrow shows the result of the final movement of MoS$_2$. (**c**) Structure of single-layer MoS$_2$ (space group P6$_3$/mmc) without strain. (**d**) Structure of single-layer MoS$_2$ (space group P6$_3$/mmc) after strain is applied. (**e**) Structure of single-layer WS$_2$ (space group R$\bar{3}$m) without strain. (**f**) Structure of single-layer WS$_2$ (space group R$\bar{3}$m) after strain is applied.

For comparison, we applied tensile strain on both WS$_2$ (space group R$\bar{3}$m) and MoS$_2$ (space group P6$_3$/mmc), which exhibited PPR. Figure 5 shows that after applying a tensile strain, the whole structure shrunk to a certain extent. Combined with the previous analysis, these two materials used as comparisons have a weaker coupling in their p-d orbitals, resulting in a reduction in the strain energy that can be stored in the bond angles during tensile strain. As a result, the changing bond angles release less energy and have a reduced effect on the surrounding atoms when the structure shrinks. The PDOS of the two compared materials did not exhibit strong antibonding properties and the atoms did not tend to repel each other, meaning they ended up exhibiting PPR. To better verify our speculation, we also calculated PDOS for four other NPR materials, and the outcomes are listed in Figure S4. It is clear from the figure that the four NPR materials have similar degrees of overlap in the positions of the peaks and energies of the PDOS, indicating that they all have strong interactions in their p-d orbitals. Such a comparison shows that the change in interatomic forces due to the interaction between p-d orbitals is a major factor in the NPR behavior of the 2D materials, which can be used as a feature for screening NPR materials. More specifically, the NPR of single-layer MoS$_2$ is related to the strong coupling between the 4d orbital of the Mo atom and the 3p orbital of the S atom, and the antibonding orbits of the atoms present. This eventually causes the interaction forces between the Mo and S atoms to tend to repel each other, leading to their extension and triggering NPR.

Finally, we performed molecular dynamics (MD) simulations at 500 K for 6 ps with NVT (Number of particles, Volume, and Temperature) ensemble to evaluate the structural stability of MoS$_2$ [49]. In the process of structural evolution, temperature fluctuations and the mean potential energy of atoms are displayed in Figure 6. Throughout the simulation process, the changes in potential energy remained near the average. Figure 6 and Figure S5 give three different configurations of MoS$_2$ and WS$_2$ structural snapshots at the end of 6 ps. The diagram shows that the geometry is well-saved and no significant structured fractures are observed, which indicates that the structure of MoS$_2$ is stable. The calculated elastic constants of the 2D molybdenum disulfide meet the Born−Huang criteria [50], in which $C_{11}, C_{22}, C_{66} > 0$ and $C_{11} + C_{22} - 2C_{12} > 0$. Judging from the data in Table S1, it is confirmed that the MoS$_2$ structure is stable.

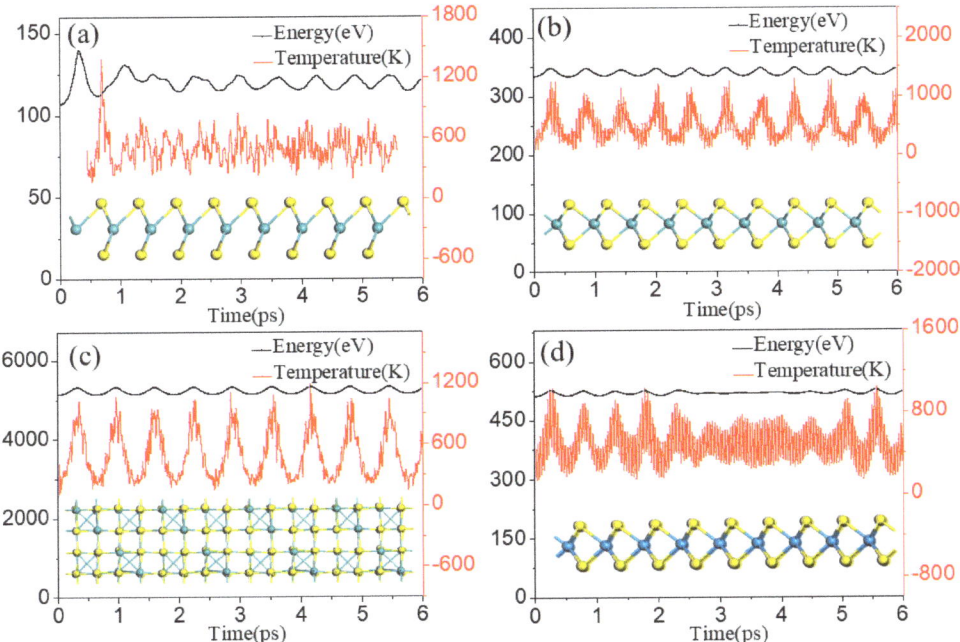

Figure 6. Structural snapshots (side view) of the (**a**) MoS$_2$ (space group R$\bar{3}$m) (**b**) MoS$_2$ (space group P6$_3$/mmc) (**c**) MoS$_2$ (space group F$\bar{4}$3m) and (**d**) WS$_2$ (space group R$\bar{3}$m), with the evolution of the mean potential energy and temperature per atom in MD (molecular dynamics) simulations at 500 K and 6 ps.

4. Conclusions

To sum up, we used first-principles calculations for the elastic constants of 2D molybdenum disulfides and have succeeded in finding a molybdenum disulfide with NPR (space group R$\bar{3}$m). This serrated structure with unique layer overlap exhibits a separate NPR of $v_x = -0.0736$ and $v_y = -0.0750$ in the X- and Y-axis. In addition, we explored the effect of layer spacing on its NPR behavior. Finally, we found that the unusual NPR behavior of single-layer MoS$_2$ is linked to its unique geometry and the strong interaction between the 4d orbitals of Mo and the 3p orbitals of S. Moreover, we also found that the strong coupling of p-d orbitals and the zigzag structure of the antibonding state are notable features for screening 2D NPR materials. These discoveries provide insight into the influence of the geometry and electronic structure of 2D molybdenum disulfide on the mechanical performance of the material. These results could propel progress in the screening and design of other 2D materials in the future.

Supplementary Materials: The following supporting information can be downloaded at https://www.mdpi.com/article/10.3390/coatings13020283/s1. Figure S1: The band structure and the DOS of MoS_2. The energy band structure describes the energy that electrons are forbidden or allowed to carry, which is caused by the diffraction of quantum dynamics electron waves in a periodic lattice. The energy band structure of a material determines various properties, especially its electronic and optical properties; Figure S2: MoS_2 (space group $F\bar{4}3m$); Figure S3: Comparison graphs of Young's modulus and Poisson's ratio for the four materials; Figure S4: Four negative Poisson's ratio materials and their PDOS plots. (a) MoS_2 (space group $P\bar{3}m1$). (b) MoS_2 (space group $I\bar{4}2d$). (c) MoSe (space group $P\bar{6}m2$). (d) WS_2 (space group $I\bar{4}2d$); Figure S5: Structural snapshots (side view) of the (a) MoS_2 (space group $R\bar{3}m$) (b) MoS_2 (space group $P6_3/mmc$) (c) MoS_2 (space group $F\bar{4}3m$) and (d) WS_2 (space group $R\bar{3}m$) with the evolution of the average potential energy and temperature per atom in AIMD simulations at 500 K and 6 ps; Table S1: The calculated elastic constants (units of GPa) of 2D MoS_2 and $MoTe_2$; and Table S2: Structural information of three configurations of MoS_2.

Author Contributions: Conceptualization, J.H.; methodology, Y.Z.; software, Y.Z.; validation, Y.Z., Y.T. and X.C.; formal analysis, Y.Z.; investigation, J.H.; resources, J.H.; data curation, Y.Z.; writing—original draft preparation, Y.Z.; writing—review and editing, Z.C.; visualization, J.H. and Y.W.; supervision, B.L. and Y.W.; project administration, B.L.; funding acquisition, B.L. All authors have read and agreed to the published version of the manuscript.

Funding: Financial support was received from the National Natural Science Foundation of China (Grant Number 21676216); the Preferential Funding Project for Scientific and Technological Activities of Overseas Chinese in Shaanxi Province (Grant Number 2021008); and the Center for High-Performance Computing of Northwestern Polytechnical University, China.

Institutional Review Board Statement: Not applicable.

Informed Consent Statement: Not applicable.

Data Availability Statement: Not applicable.

Conflicts of Interest: The authors declare that they have no known competing financial interests or personal relationships that could have appeared to influence the work reported in this paper.

References

1. Huang, C.; Zhou, J.; Wu, H.; Deng, K.; Jena, P.; Kan, E. Quantum anomalous Hall effect in ferromagnetic transition metal halides. *Phys. Rev. B* **2017**, *95*, 45113. [CrossRef]
2. Zhang, G.; Qin, G.; Zhang, F. Effects of Internal Relaxation of Biaxial Strain on Structural and Electronic Properties of $In_{0.5}Al_{0.5}N$ Thin Film. *Coatings* **2022**, *12*, 598. [CrossRef]
3. Evans, K.E.; Nkansah, M.A.; Hutchinson, I.J.; Rogers, S.C. Molecular network design. *Nature* **1991**, *353*, 124. [CrossRef]
4. Yuan, R.; Zhou, Y.; Fan, X.; Lu, Q. Negative-Poisson-Ratio polyimide aerogel fabricated by tridirectional freezing for High- and Low-Temperature and Impact-Resistant applications. *Chem. Eng. J.* **2022**, *433*, 134404. [CrossRef]
5. Yang, W.; Li, Z.; Shi, W.; Xie, B.; Yang, M. Review on auxetic materials. *J. Mater. Sci.* **2004**, *39*, 3269–3279. [CrossRef]
6. Ting, T.C.T.; Chen, T. Poisson's ratio for anisotropic elastic materials can have no bounds. *Q. J. Mech. Appl.* **2005**, *58*, 73–82. [CrossRef]
7. Hu, H.; Silberschmidt, V. A composite material with Poisson's ratio tunable from positive to negative values: An experimental and numerical study. *J. Mater. Sci.* **2013**, *48*, 8493–8500. [CrossRef]
8. Gao, Z.; Dong, X.; Li, N.; Ren, J. Novel Two-Dimensional Silicon Dioxide with in-Plane Negative Poisson's Ratio. *Nano Lett.* **2017**, *17*, 772–777. [CrossRef] [PubMed]
9. Kaminakis, N.T.; Stavroulakis, G.E. Topology optimization for compliant mechanisms, using evolutionary-hybrid algorithms and application to the design of auxetic materials. *Compos. Part B* **2012**, *43*, 2655–2668. [CrossRef]
10. Blackburn, S.; Wilson, D.L. Shaping ceramics by plastic processing. *J. Eur. Ceram. Soc.* **2008**, *7*, 1341–1351. [CrossRef]
11. Lipsett, A.W.; Beltzer, A.I. Reexamination of dynamic problems of elasticity for negative Poisson's ratio. *J. Acoust. Soc. Am.* **1988**, *84*, 2179. [CrossRef]
12. Lakes, R.; Elms, K. Indentability of Conventional and Negative Poisson's Ratio Foams. *J. Compos. Mater.* **1993**, *27*, 1193. [CrossRef]
13. Li, X.; Gao, L.; Zhou, W.; Wang, Y.; Lu, Y. Novel 2D metamaterials with negative Poisson's ratio and negative thermal expansion. *Extrem. Mech. Lett.* **2019**, *30*, 100498. [CrossRef]
14. Demir, H.; Cosgun, A.E. The Effect on Energy Efficiency of Yttria-Stabilized Zirconia on Brass, Copper and Hardened Steel Nozzle in Additive Manufacturing. *Coatings* **2022**, *12*, 690. [CrossRef]
15. Khomenko, V.; Butenko, O.; Chernysh, O.; Barsukov, V.; Suchea, M.P.; Koudoumas, E. Electromagnetic Shielding of Composite Films Based on Graphite, Graphitized Carbon Black and Iron-Oxide. *Coatings* **2022**, *12*, 665. [CrossRef]

16. Buet, E.; Braun, J.; Sauder, C. Influence of Texture and Thickness of Pyrocarbon Coatings as Interphase on the Mechanical Behavior of Specific 2.5D SiC/SiC Composites Reinforced with Hi-Nicalon S Fibers. *Coatings* **2022**, *12*, 573. [CrossRef]
17. Jiang, J.; Park, H.S. Negative Poisson's ratio in single-layer black phosphorus. *Nat. Commun.* **2014**, *5*, 4727. [CrossRef]
18. Wang, Y.; Li, F.; Li, Y.; Chen, Z. Semi-metallic Be_5C_2 monolayer global minimum with quasi-planar pentacoordinate carbons and negative Poisson's ratio. *Nat. Commun.* **2016**, *7*, 11488. [CrossRef] [PubMed]
19. Zhang, L.C.; Qin, G.; Fang, W.Z.; Cui, H.J.; Zheng, Q.R.; Yan, Q.B.; Su, G. Tinselenidene: A Two-dimensional Auxetic Material with Ultralow Lattice Thermal Conductivity and Ultrahigh Hole Mobility. *Sci. Rep.* **2016**, *6*, 19830. [CrossRef]
20. Zhou, L.; Zhuo, Z.; Kou, L.; Du, A.; Tretiak, S. Computational Dissection of Two-Dimensional Rectangular Titanium Mononitride TiN: Auxetics and Promises for Photocatalysis. *Nano Lett.* **2017**, *17*, 4466–4472. [CrossRef]
21. Qin, G.; Yan, Q.B.; Qin, Z.; Yue, S.Y.; Cui, H.J.; Zheng, Q.R.; Su, G. Hinge-like Structure Induced Unusual Properties of Black Phosphorus and New Strategies to Improve the Thermoelectric Performance. *Sci. Rep.* **2015**, *4*, 6946. [CrossRef]
22. Wang, H.; Li, X.; Li, P.; Yang, J. δ-Phosphorene: A Two-Dimensional Material with a Highly Negative Poisson's Ratio. *Nanoscale* **2017**, *9*, 850–855. [CrossRef]
23. Meng, L.B.; Ni, S.; Zhang, Y.J.; Li, B.; Zhou, X.W.; Wu, W.D. Two-Dimensional Zigzag-Shaped Cd_2C Monolayer with a Desirable Bandgap and High Carrier Mobility. *J. Mater. Chem. C* **2018**, *6*, 9175–9180. [CrossRef]
24. Meng, L.; Zhang, Y.; Zhou, M.; Zhang, J.; Zhou, X.; Ni, S.; Wu, W. Unique Zigzag-Shaped Buckling Zn_2C Monolayer with Strain-Tunable Band Gap and Negative Poisson Ratio. *Inorg. Chem.* **2018**, *57*, 1958–1963. [CrossRef]
25. Peng, R.; Ma, Y.; He, Z.; Huang, B.; Kou, L.; Dai, Y. Single-Layer Ag_2S: A Two-Dimensional Bidirectional Auxetic Semiconductor. *Nano Lett.* **2019**, *19*, 1227–1233. [CrossRef]
26. Miró, P.; Audiffred, M.; Heine, T. An atlas of two-dimensional materials. *Chem. Soc. Rev.* **2014**, *43*, 6537–6554. [CrossRef] [PubMed]
27. Jiang, L.; Marconcini, P.; Hossian, M.S.; Qiu, W.; Evans, R.; Macucci, M.; Skafidas, E. A tight binding and k.p study of monolayer stanene. *Sci. Rep.* **2017**, *7*, 12069. [CrossRef] [PubMed]
28. Yan, Y.; Yan, J.; Gong, X.; Tang, X.; Xu, X.; Meng, T.; Bu, F.; Cai, D.; Zhang, Z.; Nie, G.; et al. All-in-one asymmetric micro-supercapacitor with Negative Poisson's ratio structure based on versatile electrospun nanofibers. *Chem. Eng. J.* **2022**, *433*, 133580. [CrossRef]
29. Mak, K.F.; Lee, C.; Hone, J.; Shan, J.; Heinz, T.F. Atomically Thin MoS_2: A New Direct-Gap Semiconductor. *Phys. Rev. Lett.* **2010**, *105*, 136805. [CrossRef]
30. Yu, L.; Yan, Q.; Ruzsinszky, A. Negative Poisson's ratio in 1T-type crystalline two-dimensional transition metal dichalcogenides. *Nat. Commun.* **2017**, *8*, 15224. [CrossRef]
31. Hung, N.T.; Nugraha, A.; Saito, R. Two-dimensional MoS_2 electromechanical actuators. *J. Phys. D Appl. Phys.* **2018**, *51*, 75306. [CrossRef]
32. Esteban-Puyuelo, R.; Sarma, D.D.; Sanyal, B. Complexity of mixed allotropes of MoS_2 unraveled by first-principles theory. *Phys. Rev. B* **2020**, *102*, 165412. [CrossRef]
33. Clark, S.J.; Segall, M.D.; Pickard, C.J.; Hasnip, P.J.; Probert, M.I.J.; Refson, K.; Payne, M.C. First principles methods using CASTEP. *Z. Kristallogr.* **2005**, *220*, 567–570. [CrossRef]
34. Segall, M.D.; Lindan, P.J.D.; Probert, M.I.J.; Pickard, C.J.; Hasnip, P.J.; Clark, S.J.; Payne, M.C. First-principles simulation: Ideas, illustrations and the CASTEP code. *J. Phys. Condens. Matter* **2002**, *14*, 27117. [CrossRef]
35. Fan, Y.; Xie, D.; Ma, D.; Jing, F.; Matthews, D.T.A.; Ganesan, R.; Leng, Y. Evaluation of the Crystal Structure and Mechanical Properties of Cu Doped TiN Films. *Coatings* **2022**, *12*, 652. [CrossRef]
36. Jacobsen, E.; Lyons, R. The sliding DFT. *IEEE Signal Process. Mag.* **2003**, *20*, 74–80.
37. Perdew, J.P.; Burke, K.; Ernzerhof, M. Generalized Gradient Approximation Made Simple. *Phys. Rev. Lett.* **1996**, *77*, 3865. [CrossRef]
38. Hu, J.; Chen, W.; Zhao, X.; Su, H.; Chen, Z. Anisotropic Electronic Characteristics, Adsorption, and Stability of Low-Index $BiVO_4$ Surfaces for Photoelectrochemical Applications. *ACS Appl. Mater. Interfaces* **2018**, *10*, 5475–5484. [CrossRef] [PubMed]
39. Cadelano, E.; Palla, P.L.; Giordano, S.; Colombo, L. Elastic Properties of Hydrogenated Graphene. *Phys. Rev. B* **2010**, *82*, 235414. [CrossRef]
40. Andrew, R.C.; Mapasha, R.E.; Ukpong, A.M.; Chetty, N. Mechanical properties of graphene and boronitrene. *Phys. Rev. B* **2012**, *85*, 125428. [CrossRef]
41. Lu, C.; Chen, C. Structure-strength relations of distinct MoN phases from first-principles calculations. *Phys. Rev. Mater.* **2020**, *4*, 44002. [CrossRef]
42. Li, Y.; Wang, H.; Xie, L.; Liang, Y.; Hong, G.; Dai, H. MoS_2 Nanoparticles Grown on Graphene: An Advanced Catalyst for the Hydrogen Evolution Reaction. *J. Am. Chem. Soc.* **2011**, *19*, 7296–7299. [CrossRef]
43. Lee, C.; Wei, X.; Kysar, J.W.; Hone, J. Measurement of the Elastic Properties and Intrinsic Strength of Monolayer Graphene. *Science* **2008**, *321*, 385–388. [CrossRef]
44. Lorenz, T.; Teich, D.; Joswig, J.; Seifert, G. Theoretical Study of the Mechanical Behavior of Individual TiS_2 and MoS_2 Nanotubes. *J. Phys. Chem. C* **2012**, *21*, 11714–11721. [CrossRef]
45. Zhang, S.; Zhou, J.; Wang, Q.; Chen, X.; Kawazoe, Y.; Jena, P. Penta-graphene: A new carbon allotrope. *Proc. Natl. Acad. Sci. USA* **2015**, *112*, 2372–2377. [CrossRef]

46. Xiao, W.; Xiao, G.; Rong, Q.; Wang, L. Theoretical discovery of novel two-dimensional V^A-N binary compounds with auxiticity. *Phys. Chem. Chem. Phys.* **2018**, *20*, 22027–22037. [CrossRef] [PubMed]
47. Kou, L.; Ma, Y.; Tang, C.; Sun, Z.; Du, A.; Chen, C. Auxetic and Ferroelastic Borophane: A Novel 2D Material with Negative Poisson's Ratio and Switchable Dirac Transport Channels. *Nano Lett.* **2016**, *16*, 7910–7914. [CrossRef]
48. Jong, M.D.; Chen, W.; Angsten, T.; Jain, A.; Notestine, R.; Gamst, A.; Sluiter, M.; Ande, C.K.; Zwaag, S.V.D.; Plata, J.; et al. Charting the complete elastic properties of inorganic crystalline compounds. *Sci. Data* **2015**, *2*, 150009. [CrossRef]
49. Liu, B.; Niu, M.; Fu, J.; Xi, Z.; Lei, M.; Quhe, R. Negative Poisson's ratio in puckered two-dimensional materials. *Phys. Rev. Mater.* **2019**, *3*, 54002. [CrossRef]
50. Jin, W.; Sun, W.; Kuang, X.; Lu, C.; Kou, L. Negative Poisson Ratio in Two-Dimensional Tungsten Nitride: Synergistic Effect from Electronic and Structural Properties. *J. Phys. Chem. Lett.* **2020**, *11*, 9643–9648. [CrossRef] [PubMed]

Disclaimer/Publisher's Note: The statements, opinions and data contained in all publications are solely those of the individual author(s) and contributor(s) and not of MDPI and/or the editor(s). MDPI and/or the editor(s) disclaim responsibility for any injury to people or property resulting from any ideas, methods, instructions or products referred to in the content.

Article

Diatomite and Glucose Bioresources Jointly Synthesizing Anode/Cathode Materials for Lithium-Ion Batteries

Yun Chen [1,2], Bo Jiang [1], Yue Zhao [3], Hongbin Liu [4,*] and Tingli Ma [1,3,*]

1. Graduate School of Life Science and Systems Engineering, Kyushu Institute of Technology, 2-4 Hibikino, Wakamatsu, Kitakyushu 808-0196, Japan
2. Medical Engineering and Technology Research Center, School of Radiology, Shandong First Medical University & Shandong Academy of Medical Sciences, Taian 271000, China
3. College of Materials and Chemistry, China Jiliang University, Hangzhou 310018, China
4. School of Materials Science and Engineering, Harbin Institute of Technology (Shenzhen), Shenzhen 518055, China
* Correspondence: liuhongbin@hit.edu.cn (H.L.); tinglima@life.kyutech.ac.jp (T.M.)

Abstract: Large-scale popularization and application make the role of lithium-ion batteries increasingly prominent and the requirements for energy density have increased significantly. The silicon-based material has ultra-high specific capacity, which is expected in the construction of next-generation high specific-energy batteries. In order to improve conductivity and maintain structural stability of the silicon anode in application, and further improve the energy density of the lithium-ion battery, we designed and synthesized carbon-coated porous silicon structures using diatomite and polysaccharides as raw materials. The electrode materials constructed of diatomite exhibit porous structures, which can provide fast transport channels for lithium ions, and effectively release the stress caused by volume expansion during cycling. At the same time, the electrical conductivity of the materials has been significantly improved by compounding with biomass carbon, so the batteries exhibit stable electrochemical performance. We systematically studied the effect of different contents of biomass carbon on the Li_2MnSiO_4/C cathode, and the results showed that the carbon content of 20% exhibited the best electrochemical performance. At a current density of 0.05C, the capacity close to 150 mAh g^{-1} can be obtained after 50 cycles, which is more than three times that of without biomass carbon. The silicon-based anode composited with biomass carbon also showed excellent cycle stability; it could still have a specific capacity of 1063 mAh g^{-1} after 100 cycles at the current density of 0.1 A g^{-1}. This study sheds light on a way of synthesizing high specific-capacity electrode materials of the lithium-ion battery from natural raw materials.

Keywords: diatomite; lithium-ion batteries; Li_2MnSiO_4/C cathode; natural raw materials

1. Introduction

With the development of modern industry and human society, fossil energy is depleted day by day, and environmental issues have received increasing attention. Researchers have begun to turn their attention to renewable and clean energy such as solar energy, tidal energy, and wind energy, etc. Secondary batteries, fuel cells, and supercapacitors are the most common high-efficiency energy storage devices [1–3]. Among these, secondary batteries stand out due to their high energy density, strong environmental adaptability, and wide application range [4,5]. In recent years, lithium-ion batteries have been widely used in portable electronic devices, electric vehicles, household power supply, etc. With the popularity of electric vehicles, the demand for lithium-ion batteries has particularly surged. Therefore, lithium-ion batteries with higher energy density, lower cost, and stable operation are the source of high hopes [6]. At the same time, as an important part of the battery, the synthesis of electrode materials with high, specific energy and low cost is an important area of research [7].

In recent years, researchers have begun to use bio-renewable and sustainable source raw materials to synthesize electrode materials for reducing environmental damage caused by waste [8–10]. Before that, biomass materials with multifarious and abundant resources have been widely used in electrical energy storage, electrocatalysis, photocatalysis, multiphase catalysis, bio-fuel environment, etc. [11,12]. Currently, biomass and biowaste-based carbon materials as anode materials for lithium-ion batteries have received extensive attention due to the advantages of being inexpensive, abundant, and environmentally friendly [13,14]. The basic elements of carbon, sulfur, nitrogen, and phosphorus can increase the wettability and reduce the transfer resistance of the biomass-derived carbon material, which contributes to the increase in battery capacity [15]. Meanwhile, biomass-derived carbon materials possess naturally ordered unique hierarchical structures as well as abundant surface properties and active sites, which facilitate ion transfer and diffusion [16,17]. Thus, research has been conducted on the use of low-cost, environmentally friendly biomass materials to synthesize electrode materials for lithium-ion batteries [18]. The electrode materials that have been reported to use biomass synthesis include rice husks, bamboo leaves, diatomite, etc. [19,20].

Diatomite is a bio-deposited siliceous rock formed from diatom remains over a long period of time in the natural environment; it is mainly composed of amorphous protein minerals of SiO_2 [21,22]. Diatomite is non-toxic, high in purity, and low in acquisition cost. Natural diatoms have different shapes, resulting in the formation of diatomite showing a variety of different microscopic forms, mostly in the shape of a round sieve, a column, or a belt [23–25]. The wall shell of diatomite is composed of multi-level, large, and orderly arranged micropores which are widely used as reaction catalysts, fillers, thermal insulation materials, and filter materials due to the properties of being lightweight, porous, with a large specific surface area, and a strong adsorption capacity [26,27].

Among the electrode materials of lithium-ion batteries, Li_2MnSiO_4 (LMS) cathode electrode material and Si anode electrode material have been reported with high energy density [28,29]. For cathode materials, Li_2MnSiO_4 has high theoretical capacitance (333 mA hg^{-1}) and excellent thermal stability due to the Si–O covalent bond [30,31]. Silicon is one of the most valuable anode materials of lithium-ion battery because it has a high theoretical capacity (~4200 mA hg^{-1}), low working potential, is an abundant resource, and is environmentally friendly [32–34]. Both of these materials can be synthesized using diatomite, so the cost can be effectively reduced. However, the silicon anode electrode has a large volume of more than 400% due to charging and discharging. Therefore, it is necessary to modify the material in a suitable way in order to bring out the advantages of the silicon anode [35–37]. Researchers have explored a variety of strategies to solve the issues, such as designing the structure of silicon materials which contain nano-silicon, layered silicon, porous silicon, etc. Another more effective way is to compound silicon with other materials to maintain the stability of silicon in the cycle, which contains carbon, metal oxide, sulfide, two-dimensional materials, etc. [38–40]. Glucose is a good carbon source that can build a three-dimensional conductive carbon network inside the materials, further improving the electronic conductivity of the materials [41,42].

In this work, we synthesized a diatomite-derived Li_2MnSiO_4/C cathode electrode material and a Si/C anode electrode material with excellent electrochemical performance. New smaller holes could be generated between the diatomaceous earth particles during the preparation of the Si material. In addition, the fabricated anode material maintained a discharge capacity of 1063 mAh g^{-1} after 100 cycles, showing excellent electrical performance and good stability. While the expansion and increase in internal resistance of the anode material can be suppressed, both the capacity retention rate and the coulombic efficiency can be improved. The Li_2MnSiO_4/C cathode also delivered a specific capacity of 249.3 mAh g^{-1} in the first cycle and maintained a specific capacity of about 150 mAh g^{-1} after 50 cycles were compared. The results show that the synthesized Li_2MnSiO_4/C cathode material has good electrochemical performance and stability. For most lithium electrode materials, although a high capacity can be achieved, the extraction process is highly complicated and costly. We used low-cost diatomite as a precursor to synthesize lithium electrode materials and obtained a higher theoretical capacity for the first time. This study provides a strong candidate for the utilization of environmentally friendly biomass towards efficient energy storage through facile and low-cost procedures.

2. Materials and Methods

All chemicals were of analytical grade and used as received without further purification. Diatomite (Wako, 99%), $MnCO_3 \cdot nH_2O$ (Wako, 99%), $LiOH \cdot H_2O$ (Wako, 99%), Mg (Wako, 99%), and NaCl (Wako, 99%).

2.1. Fabrication of Cathode Electrode Li_2MnSiO_4/C

Diatomite was treated with 1M HCl solution to obtain high purity SiO_2. The obtained SiO_2, $MnCO_3 \cdot nH_2O$, and $LiOH \cdot H_2O$ were individually, finely pulverized in an agate mortar for 30 min. Then, the various processed materials were mixed together and ground for another 15 min. Next, the mixed sample was added to the ethanol solution and stirred thoroughly for dispersion for 6 h. After vacuum drying, calcining in tube furnace for 8 h under argon atmosphere at 700 °C, the Li_2MnSiO_4 material was obtained. Finally, the glucose was added to Li_2MnSiO_4 material to obtain different proportions of carbon (weight ratio 0%, 10%, 20%) by calcining in tube furnace for 2 h under argon atmosphere at 600 °C to prepare three cathode materials of LMS/wt.0% C, LMS/wt.10% C, and LMS/wt.20% C.

2.2. Fabrication of Anode Electrode Si/C

First, the treated diatomite, Mg powder, and NaCl were ground in an agate mortar for 30 min. The mixing sample was calcined in the tube furnace for 8 h under argon atmosphere at 700 °C [43]. Then, the argon was replaced with nitrogen and calcined for another 8 h. Next, the obtained sample was added to 1M HCl solution to remove impurities to obtain porous silicon material. After compounding with glucose, carbonizing for 2 h under argon atmosphere at 600 °C, the final Si/C anode material with carbon content of 10% was obtained (DS–3). The additional samples DS–1 (without NaCl and N_2 process) and DS–2 (without N_2 process) were also prepared according to different conditions.

The synthesis procedure is shown in Figure 1.

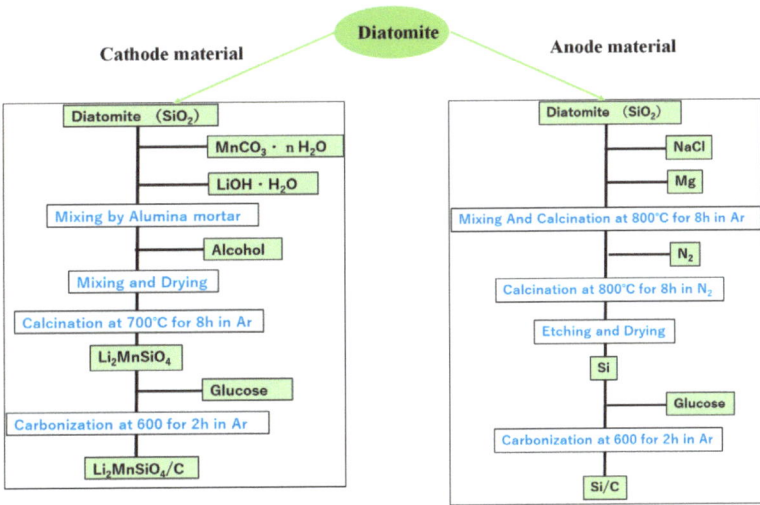

Figure 1. Synthesis process of Li_2MnSiO_4/C and Si/C materials.

2.3. Fabrication of Coin Batteries

The active materials, conductive additive acetylene black, and polyvinylidene difluoride (PVDF) were mixed with NMP at a ratio of 80:10:10 to form homogenous slurries. The slurries were coated on aluminum foil for cathode material and copper foil for anode material. The obtained electrodes were dried at 90 °C for 2 h, and then transferred to vacuum oven at 120 °C for 10 hours. The mass loading of active materials in the working electrodes was 1.0–1.5 mg/cm^2, with lithium metal as the counter electrode. The 2032 type button batteries were manufactured using the manufactured electrodes. The microporous polypropylene membrane was used as a separator. The electrolyte was 1 M LiPF$_6$ in a mixed ethylene carbonate/diethyl carbonate solvent (1:1) with 5% fluoroethylene carbonate (FEC) additive.

3. Results and Discussion

3.1. Morphology Characteristics

The scanning electron microscope (SEM) measurement was used to observe and analyze the characteristics of material morphology, as shown in Figure 2. The diatomite we used has a complete shape, and the micropores on the wall are uniform in size and arranged in an orderly manner. It is an exquisite natural porous material (Figure 2a). The diatomite in the shape of a round cake has a rich and well-developed honeycomb porous structure (Figure 2b). Thus, the large specific surface area of diatomite is conducive to the embedding of lithium ions, and the porous circular cake structure can provide more buffer space to enhance the stability. The Si/C anode material obtained by reducing diatomite maintains the characteristics, and the surface is rougher from the increased porous structure (Figure 2c,d). This special structure can maintain the structural stability of the Si-based material during cycling, which was also proved by subsequent electrochemical tests. The surface morphology of Li$_2$MnSiO$_4$/C cathode material is different from the original diatomite. The observed void structure is significantly smaller, and the surface has obvious granularity (Figure 2e,f). This is mainly because the synthesis process of Li$_2$MnSiO$_4$/C cathode material is an "addition" reaction, which is different from the diatomite-based "reduction" reaction of the Si/C anode material in the reaction process.

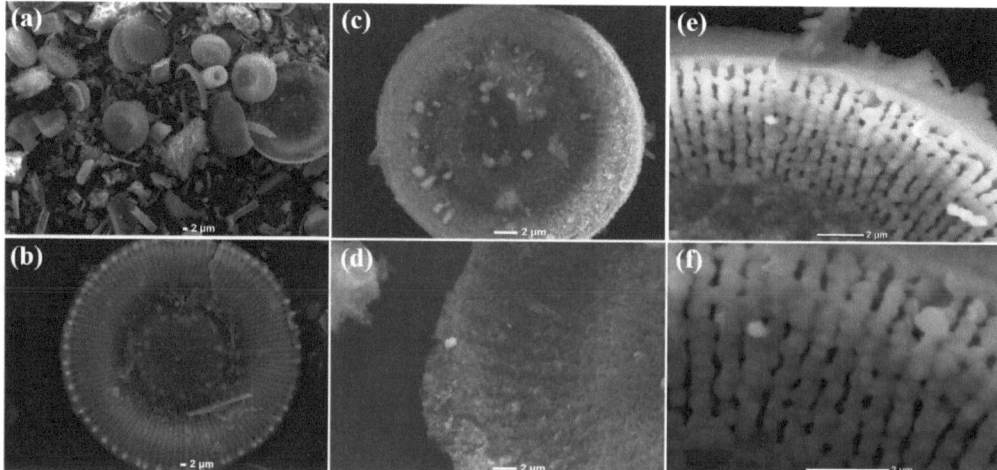

Figure 2. The SEM images of (**a**,**b**) diatomite, (**c**,**d**) Si/C anode material, and (**e**,**f**) Li$_2$MnSiO$_4$/C cathode material.

Transmission electron microscopy (TEM) and energy-dispersive X-ray spectroscopy (EDX) were used to further analyze the structure of the samples on a smaller scale, which is shown in Figures 3 and 4. It shows that the Li$_2$MnSiO$_4$/C cathode material is composed of

100 nm~200 nm particles (Figure 3a,b). EDX element mapping displays that the elements of O, Si, and Mn are evenly distributed on the surface of the Li$_2$MnSiO$_4$/C composite (Figure 3c–e). And the Si/C anode material is distributed with a porous structure and with Si elements uniformly distributed (Figure 4).

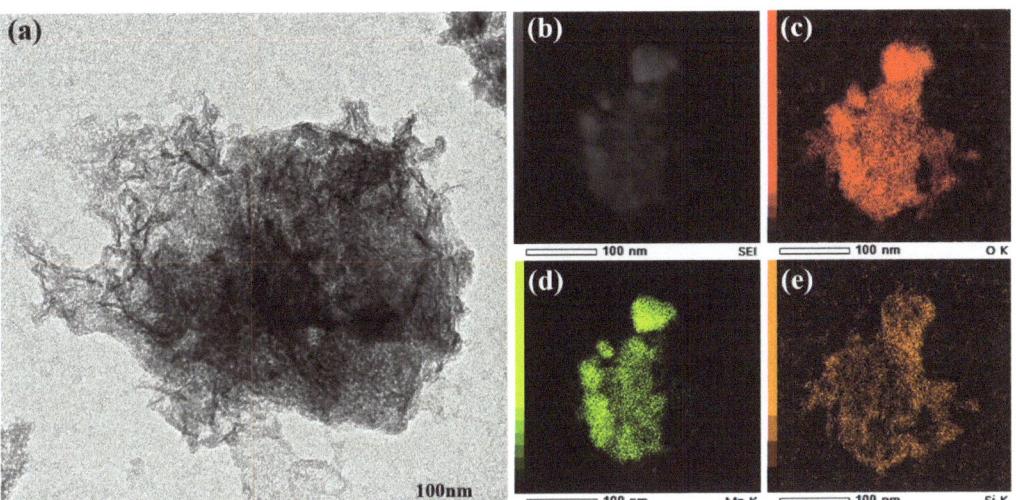

Figure 3. The (**a**,**b**) TEM images of Li$_2$MnSiO$_4$/C cathode material; the EDX mapping images: (**c**) O, (**d**) Mn, and (**e**) Si elements distribution.

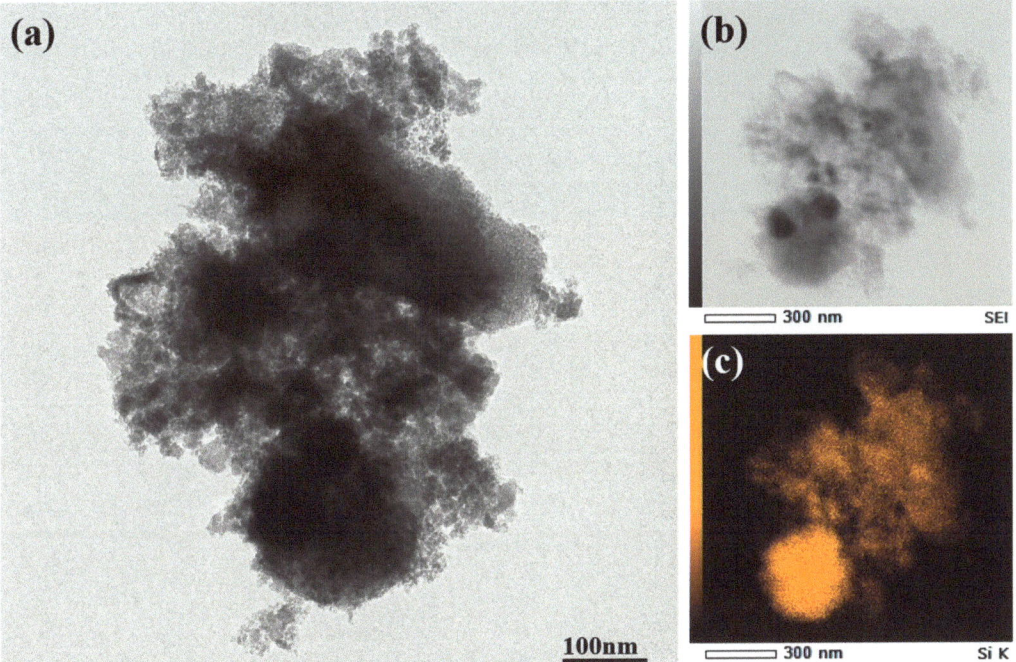

Figure 4. The (**a**,**b**) TEM images and (**c**) EDX mapping image of Si/C anode material.

3.2. Structural Characteristics

The X-ray diffraction (XRD) patterns were used to determine the main composition of the Li_2MnSiO_4/C cathode and Si/C anode. It was found that the XRD pattern of Li_2MnSiO_4/C contained an impure phase that belonged to MnO as shown in Figure 5a. However, other characteristic peaks are consistent with the pure Li_2MnSiO_4 phase. Figure 5b shows the XRD diffraction patterns of different Si/C anode materials prepared by changing the synthesis conditions. The characteristic peaks at 76.7°, 69°, 56.2°, 47.4°, and 28.5° are attributed to the (331), (400), (311), (220), and (111) of pure silicon lattice planes, respectively [34]. Therefore, it was confirmed that there are no other impurities in the synthesized samples.

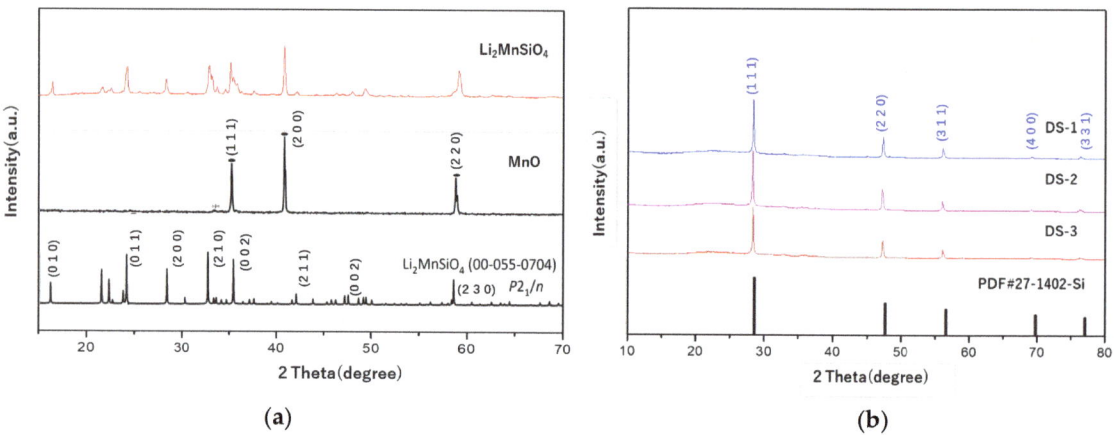

Figure 5. (**a**) The XRD patterns of Li_2MnSiO_4/C cathode and (**b**) Si/C anode.

The specific surface area and pore volume of the samples were measured by the N_2 adsorption method based on Brunauer–Emmett–Teller (BET), which is shown in Figure 6 and Table 1. The diatomite has a specific surface area of 32.6 m^2 g^{-1}, while the specific surface area of the Li_2MnSiO_4/C cathode material is reduced by about five times to 6.4 m^2 g^{-1} (Figure 6a). Due to the generated internal "addition" reaction with the $MnCO_3 \cdot nH_2O$ and $LiOH \cdot H_2O$, the mass of diatomite increases, which will significantly reduce the size of the structural pores, which is also confirmed by the SEM. Different from the Li_2MnSiO_4/C cathode material, the specific surface areas of the synthesized anode materials DS–1, DS–2, and DS–3 increase to 137.6, 160.6, and 253.0 m^2 g^{-1}, respectively. As a result of the "reduction" reaction, the oxygen atoms in the diatomite are released, and more space is reserved in the structure.

Table 1. BET measurement result of materials.

	Material	Specific Surface Area/m^2 g^{-1}	Pore Volume/m^3 g^{-1}
Precursor	Diatomite	32.6	0.05
Cathode	Li_2MnSiO_4/C	6.4	0.01
Anode	DS–1	137.6	0.18
	DS–2	160.6	0.56
	DS–3	253.0	1.07

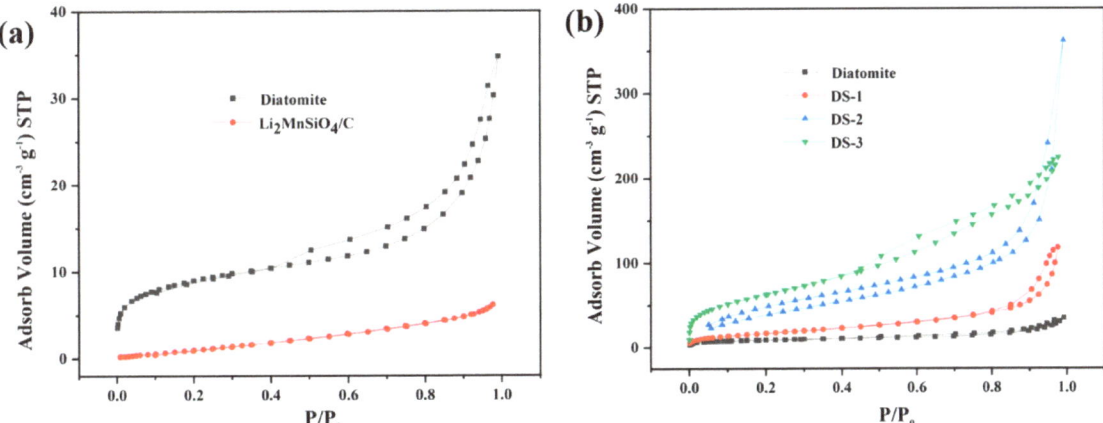

Figure 6. The nitrogen adsorption–desorption isotherms of (**a**) diatomite and Li$_2$MnSiO$_4$/C; and (**b**) SD−1, DS−2, and DS−3.

3.3. Electrochemical Performance

The cyclic voltammetry (CV) curves of the Li$_2$MnSiO$_4$/C cathode and the DS-3 anode are shown in Figure 7. For Li$_2$MnSiO$_4$/C cathode, the peaks around 3.6 V and 4.3 V at the cathodic scan are due to the conversion reaction of Mn^{2+} to Mn^{3+} and Mn^{4+}. The peaks at 2.0 V, 2.8 V, and 4.4 V at the cathodic scan are due to the reversible reduction of Mn^{4+} to Mn^{2+} (Figure 7a). Figure 7b shows the redox curve of a typical Si-based anode. The cathodic peak at 0.1 V corresponds to the formation of an amorphous Si–Li alloy, while two anodic peaks at 0.25 and 0.5 V are due to the conversion reaction of Li$_x$Si to amorphous Si. Figure 7c shows the charge–discharge curves of Li$_2$MnSiO$_4$/C with different carbon content during the first cycle at a current density of 0.05C. The discharge capacity of LMS/wt.0% C is 187.4 mAhg^{-1}, which is 59% of the charge capacity of 318.2 mAhg^{-1}, while the LMS/wt.10% C and LMS/wt.20% cathode exhibit over 80% initial coulomb efficiency. Figure 7d shows the cycle curves of Li$_2$MnSiO$_4$/C cathode materials at a current density of 0.05C. Compared to LMS/wt.0% C and LMS/wt.10% C, LMS/wt.20% C has the best electrochemical performance. For LMS/wt.20% C, the specific capacity is about 150 mAh g^{-1} after 50 cycles, which proved it has stable cycling performance. The results of LMS/wt.10% C and LMS/wt.20% proved that combining with carbon is an effective way to improve the conductivity of the Li$_2$MnSiO$_4$ cathode. Figure 7e shows the cycle curves of the Si/C anode. After 100 cycles, the discharge specific capacity of the DS-3 anode is 1063 mA h g^{-1}, which is much superior to DS-1 and DS-2. This is due to the fact that the original shape and porous structure of diatomite are maintained during the synthesis process. In addition, the participation of N$_2$ is conducive to the generation of porous structure which provides more buffer space for the expansion of the Si-based anode during cycling. At the same time, the rate capability test is also used to confirm the performance of Si/C anode materials at different current densities (Figure 7f). As expected, the DS-3 anode material exhibits the best rate performance, and when the current density returned to 100 mA g^{-1}, its specific capacity returned to above 1000 mAh g^{-1}.

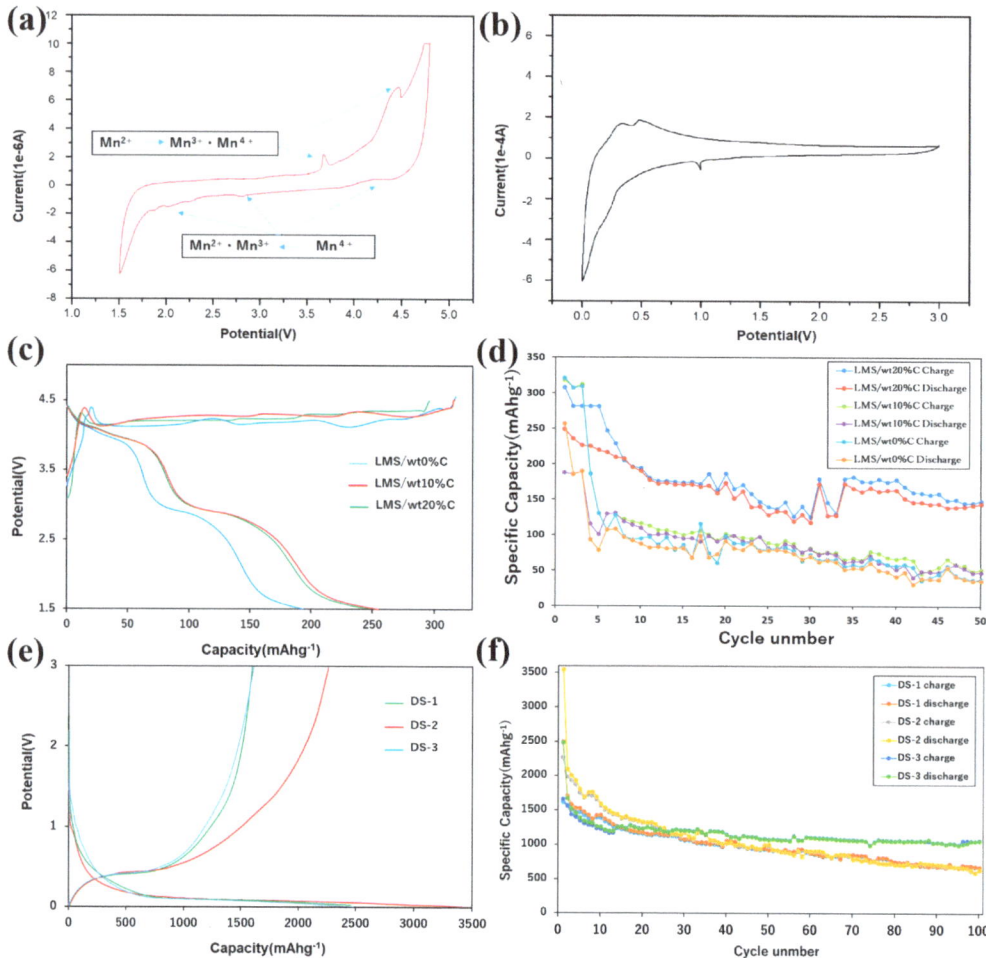

Figure 7. The CV curves of (**a**) Li$_2$MnSiO$_4$/C cathode under the potential range from 1.5 to 4.8 V and (**b**) DS−3 anode under the potential range from 0.005 to 3.0 V at 0.1 mV/s. (**c**) Charge/discharge curve diagram and (**d**) cycling performance of Li$_2$MnSiO$_4$ cathode under the potential range from 1.5 to 4.5 V at a current density of 0.05C. (**e**) Charge/discharge curve diagram and (**f**) cycling performance of Si/C anode materials under the potential range from 0.005 to 3.0 V at a current density of 0.1 A g^{-1}.

4. Conclusions

In this work, we designed and synthesized a porous Li$_2$MnSiO$_4$/C cathode and Si/C anode materials using natural diatomite and glucose. By compounding, the electrical conductivity of the materials is significantly improved, and the porous structures of the materials can effectively maintain stability. Through gradient comparison, the Li$_2$MnSiO$_4$/C cathode material prepared with 20% glucose has better electrochemical performance and capacity retention after 50 cycles. With a simple mixed calcination treatment, we obtained a stable Si/C anode material with high specific capacity. After different conditions and comparisons, we screened out the best reaction conditions. By mixing NaCl and adding a step of nitrogen calcination, the treated materials react more fully, the impurities can be effectively removed, and a purer product sample DS-3 can be obtained which has a smaller particle size distribution. It has a larger specific surface area of 253 m^2 g^{-1} while still maintaining a porous structure, which has a positive effect on the fast transport of lithium ions. Thus, the cycling performance of the material can be improved. After 100 cycles, the material can maintain the capacity of above 1000 mAh g^{-1} at the current density of 100 mA g^{-1}. This provides a very good reference for the use of natural materials to construct high specific-energy lithium-ion batteries.

Author Contributions: Methodology, B.J.; Software, Y.Z.; Investigation, Y.C. and B.J.; Data curation, Y.Z.; Writing—original draft, Y.C., H.L. and T.M.; Writing—review & editing, H.L. and T.M. All authors have read and agreed to the published version of the manuscript.

Funding: This work was conducted at Kitakyushu Foundation for the Advancement of Industry, Science and Technology, Semiconductor Center, and supported by "Nanotechnology Platform Program" of the Ministry of Education, Culture, Sports, Science and Technology (MEXT), Japan, Grant Number JPMXP09F-19-FA-0029 and KAGENHI Grant-in-Aid for Scientific Research(B), No.19H02818, and the National Natural Science Foundation of China (Grant No. 51772039 and 51972293).

Institutional Review Board Statement: Not applicable.

Informed Consent Statement: Not applicable.

Data Availability Statement: Not applicable.

Conflicts of Interest: The authors declare no conflict of interest.

References

1. Duffner, F.; Kronemeyer, N.; Tübke, J.; Leker, J.; Winter, M.; Schmuch, R. Post-lithium-ion battery cell production and its compatibility with lithium-ion cell production infrastructure. *Nat. Energy* **2021**, *6*, 123–134. [CrossRef]
2. Xiong, R.; Pan, Y.; Shen, W.; Li, H.; Sun, F. Lithium-ion battery aging mechanisms and diagnosis method for automotive applications: Recent advances and perspectives. *Renew. Sustain. Energy Rev.* **2020**, *131*, 110048. [CrossRef]
3. Sasrimuang, S.; Chuchuen, O.; Artnaseaw, A. Synthesis, characterization, and electrochemical properties of carbon nanotubes used as cathode materials for Al–air batteries from a renewable source of water hyacinth. *Green Process. Synth.* **2020**, *9*, 340–348. [CrossRef]
4. Lyu, P.; Liu, X.; Qu, J.; Zhao, J.; Huo, Y.; Qu, Z.; Rao, Z. Recent advances of thermal safety of lithium ion battery for energy storage. *Energy Storage Mater.* **2020**, *31*, 195–220. [CrossRef]
5. Pushnitsa, K.; Kosenko, A.; Chernyavsky, V.; Pavlovskii, A.A.; Novikov, P.; Popovich, A.A. Copper-Coated Graphite Felt as Current Collector for Li-Ion Batteries. *Coatings* **2022**, *12*, 1321. [CrossRef]
6. Sommerville, R.; Zhu, P.; Rajaeifar, M.A.; Heidrich, O.; Goodship, V.; Kendrick, E. A qualitative assessment of lithium ion battery recycling processes. *Resour. Conserv. Recycl.* **2021**, *165*, 105219. [CrossRef]
7. Liu, K.; Wei, Z.; Yang, Z.; Li, K. Mass load prediction for lithium-ion battery electrode clean production: A machine learning approach. *J. Clean. Prod.* **2021**, *289*, 125159. [CrossRef]
8. Chen, Y.; Liu, H.; Jiang, B.; Zhao, Y.; Meng, X.; Ma, T. Hierarchical porous architectures derived from low-cost biomass equisetum arvense as a promising anode material for lithium-ion batteries. *J. Mol. Struct.* **2020**, *1221*, 128794. [CrossRef]
9. Zhang, Y.; Zhang, R.; Chen, S.; Gao, H.; Li, M.; Song, X.; Xin, H.L.; Chen, Z. Diatomite-Derived Hierarchical Porous Crystalline-AmorphousNetwork for High-Performance and Sustainable Si Anodes. *Adv. Funct. Mater.* **2020**, *30*, 2005956. [CrossRef]
10. Wu, W.; Wang, M.; Wang, J.; Wang, C.; Deng, Y. Green Design of Si/SiO$_2$/C Composites as High-Performance Anodes for Lithium-Ion Batteries. *ACS Appl. Energy Mater.* **2020**, *3*, 3884–3892. [CrossRef]
11. Shibutani, N.; Sugiura, T.; Tanaka, A.; Nagato, K. Examination on Water Management Method in the Same Electrode in PEFC. *ECS Trans.* **2021**, *104*, 243. [CrossRef]
12. Ribeiro, M.J.; Tulyaganov, D.U.; Ferreira, J.M.F.; Labrincha, J.A. Production of Al-rich sludge-containing ceramic bodies by different shaping techniques. *J. Mater. Process. Technol.* **2004**, *148*, 139–146. [CrossRef]
13. Nie, W.; Cheng, H.; Liu, X.; Sun, Q.; Tian, F.; Yao, W.; Liang, S.; Lu, X.; Zhou, J. Surface organic nitrogen-doping disordered biomass carbon materials with superior cycle stability in the sodium-ion batteries. *J. Power Sources* **2022**, *522*, 230994. [CrossRef]
14. Dominko, R.; Bele, M.; Gaberšček, M.; Meden, A.; Remškar, M.; Jamnik, J. Structure and electrochemical performance of Li$_2$MnSiO$_4$ and Li$_2$FeSiO$_4$ as potential Li-battery cathode materials. *Electrochem. Commun.* **2006**, *8*, 217–222. [CrossRef]
15. Wang, P.; Chen, L.; Shen, Y. Recycling spent ternary lithium-ion batteries for modification of dolomite used in catalytic biomass pyrolysis–A preliminary study by thermogravimetric and pyrolysis-gas chromatography/mass spectrometry analysis. *Bioresour. Technol.* **2021**, *337*, 125476. [CrossRef]
16. Yu, J.; Tang, T.; Cheng, F.; Huang, D.; Martin, J.L.; Brewer, C.E.; Grimm, R.L.; Zhou, M.; Luo, H. Exploring spent biomass-derived adsorbents as anodes for lithium ion batteries. *Mater. Today Energy* **2021**, *19*, 100580. [CrossRef]
17. Ma, Q.; Dai, Y.; Wang, H.; Ma, G.; Guo, H.; Zeng, X.; Tu, N.; Wu, X.; Xiao, M. Directly conversion the biomass-waste to Si/C composite anode materials for advanced lithium ion batteries. *Chin. Chem. Lett.* **2021**, *32*, 5–8. [CrossRef]
18. Vernardou, D. Recent Report on the Hydrothermal Growth of LiFePO4 as a Cathode Material. *Coatings* **2022**, *12*, 1543. [CrossRef]

19. Guo, X.; Zheng, S.; Luo, Y.; Pang, H. Synthesis of confining cobalt nanoparticles within SiO_x/nitrogen-doped carbon framework derived from sustainable bamboo leaves as oxygen electrocatalysts for rechargeable Zn-air batteries. *Chem. Eng. J.* **2020**, *401*, 126005. [CrossRef]
20. Chen, Y.; Guo, X.; Liu, A.; Zhu, H.; Ma, T. Recent progress in biomass-derived carbon materials used for secondary batteries. *Sustain. Energy Fuels* **2021**, *5*, 3017–3038. [CrossRef]
21. Bakr, H. Diatomite: Its characterization, modifications and applications. *Asian J. Mater. Sci.* **2010**, *2*, 121–136.
22. Zhou, F.; Li, Z.; Lu, Y.-Y.; Shen, B.; Guan, Y.; Wang, X.-X.; Yin, Y.-C.; Zhu, B.-S.; Lu, L.-L.; Ni, Y.; et al. Diatomite derived hierarchical hybrid anode for high performance all-solid-state lithium metal batteries. *Nat. Commun.* **2019**, *10*, 2482. [CrossRef]
23. Ivanov, S.É.; Belyakov, A. Diatomite and its applications. *Glass Ceram.* **2008**, *65*, 18–21. [CrossRef]
24. Chen, X.; Tian, Y. Review of Graphene in Cathode Materials for Lithium-Ion Batteries. *Energy Fuels* **2021**, *35*, 3572–3580. [CrossRef]
25. Kang, M.S.; Heo, I.; Kim, S.; Yang, J.; Kim, J.; Min, S.-J.; Chae, J.; Yoo, W.C. High-areal-capacity of micron-sized silicon anodes in lithium-ion batteries by using wrinkled-multilayered-graphenes. *Energy Storage Mater.* **2022**, *50*, 234–242. [CrossRef]
26. de Namor, A.F.D.; El Gamouz, A.; Frangie, S.; Martinez, V.; Valiente, L.; Webb, O.A. Turning the volume down on heavy metals using tuned diatomite. A review of diatomite and modified diatomite for the extraction of heavy metals from water. *J. Hazard. Mater.* **2012**, *241*, 14–31. [CrossRef]
27. Caliskan, N.; Kul, A.R.; Alkan, S.; Sogut, E.G.; Alacabey, I. Adsorption of Zinc (II) on diatomite and manganese-oxide-modified diatomite: A kinetic and equilibrium study. *J. Hazard. Mater.* **2011**, *193*, 27–36. [CrossRef] [PubMed]
28. Cheng, Q.; He, W.; Zhang, X.; Li, M.; Wang, L. Modification of Li_2MnSiO_4 cathode materials for lithium-ion batteries: A review. *J. Mater. Chem. A* **2017**, *5*, 10772–10797. [CrossRef]
29. Shree Kesavan, K.; Michael, M.S.; Prabaharan, S.R.S. Facile Electrochemical Activity of Monoclinic Li_2MnSiO_4 as Potential Cathode for Li-Ion Batteries. *ACS Appl. Mater. Interfaces* **2019**, *11*, 28868–28877. [CrossRef]
30. Singh, M.; Kumar, N.; Sharma, Y. Role of impurity phases present in orthorhombic-Li_2MnSiO_4 towards the Li-reactivity and storage as LIB cathode. *Appl. Surf. Sci.* **2022**, *574*, 151689. [CrossRef]
31. Wu, X.; Zhao, S.-X.; Yu, L.-Q.; Yang, J.-L.; Nan, C.-W. Effect of sulfur doping on structural reversibility and cycling stability of a Li_2MnSiO_4 cathode material. *Dalton Trans.* **2018**, *47*, 12337–12344. [CrossRef]
32. Zhu, P.; Gastol, D.; Marshall, J.; Sommerville, R.; Goodship, V.; Kendrick, E. A review of current collectors for lithium-ion batteries. *J. Power Sources* **2021**, *485*, 229321. [CrossRef]
33. Li, J.; Xu, J.; Xie, Z.; Gao, X.; Zhou, J.; Xiong, Y.; Chen, C.; Zhang, J.; Liu, Z. Diatomite-Templated Synthesis of Freestanding 3D Graphdiyne for Energy Storage and Catalysis Application. *Adv. Mater.* **2018**, *30*, 1800548. [CrossRef]
34. Liu, H.; Meng, X.; Chen, Y.; Zhao, Y.; Guo, X.; Ma, T. Synthesis and Surface Engineering of Composite Anodes by Coating Thin-Layer Silicon on Carbon Cloth for Lithium Storage with High Stability and Performance. *ACS Appl. Energy Mater.* **2021**, *4*, 6982–6990. [CrossRef]
35. Liu, Z.; Lu, D.; Wang, W.; Yue, L.; Zhu, J.; Zhao, L.; Zheng, H.; Wang, J.; Li, Y. Integrating Dually Encapsulated Si Architecture and Dense Structural Engineering for Ultrahigh Volumetric and Areal Capacity of Lithium Storage. *ACS Nano* **2022**, *16*, 4642–4653. [CrossRef] [PubMed]
36. Tong, L.; Long, K.; Chen, L.; Wu, Z.; Chen, Y. High-Capacity and Long-Lived Silicon Anodes Enabled by Three-Dimensional Porous Conductive Network Design and Surface Reconstruction. *ACS Appl. Energy Mater.* **2022**, *5*, 13877–13886. [CrossRef]
37. Zheng, P.; Sun, J.; Liu, H.; Wang, R.; Liu, C.; Zhao, Y.; Li, J.; Zheng, Y.; Rui, X. Microstructure Engineered Silicon Alloy Anodes for Lithium-Ion Batteries: Advances and Challenges. *Batter. Supercaps* **2022**.
38. Kim, J.; Park, Y.K.; Kim, H.; Jung, I.H. Ambidextrous Polymeric Binder for Silicon Anodes in Lithium-Ion Batteries. *Chem. Mater.* **2022**, *34*, 5791–5798. [CrossRef]
39. Li, L.; Li, T.; Sha, Y.; Ren, B.; Zhang, L.; Zhang, S. A Web-like Three-dimensional Binder for Silicon Anode in Lithium-ion Batteries. *Energy Environ. Mater.* **2022**, e12482. [CrossRef]
40. Liu, P.; Li, B.; Zhang, J.; Jiang, H.; Su, Z.; Lai, C. Self-swelling derived frameworks with rigidity and flexibility enabling high-reversible silicon anodes. *Chin. Chem. Lett.* **2022**, 107946. [CrossRef]
41. Bai, J.; Zhao, B.; Zhou, J.; Si, J.; Fang, Z.; Li, K.; Ma, H.; Dai, J.; Zhu, X.; Sun, Y. Glucose-induced synthesis of 1T-MoS2/C hybrid for high-rate lithium-ion batteries. *Small* **2019**, *15*, 1805420. [CrossRef] [PubMed]
42. Liu, H.; Luo, S.; Yan, S.; Wang, Q.; Hu, D.; Wang, Y.; Feng, J.; Yi, T. High-performance α-Fe_2O_3/C composite anodes for lithium-ion batteries synthesized by hydrothermal carbonization glucose method used pickled iron oxide red as raw material. *Compos. Part B Eng.* **2019**, *164*, 576–582. [CrossRef]
43. Yoon, T.; Bok, T.; Kim, C.; Na, Y.; Park, S.; Kim, K.S. Mesoporous Silicon Hollow Nanocubes Derived from Metal–Organic Framework Template for Advanced Lithium-Ion Battery Anode. *ACS Nano* **2017**, *11*, 4808–4815. [CrossRef] [PubMed]

Disclaimer/Publisher's Note: The statements, opinions and data contained in all publications are solely those of the individual author(s) and contributor(s) and not of MDPI and/or the editor(s). MDPI and/or the editor(s) disclaim responsibility for any injury to people or property resulting from any ideas, methods, instructions or products referred to in the content.

Article

Study of Co-Deposition of Tantalum and Titanium during the Formation of Layered Composite Materials by Magnetron Sputtering

Elena Olegovna Nasakina *, Maria Andreevna Sudarchikova, Konstantin Yurievich Demin, Alexandra Borisovna Mikhailova, Konstantin Vladimirovich Sergienko, Sergey Viktorovich Konushkin, Mikhail Alexandrovich Kaplan, Alexander Sergeevich Baikin, Mikhail Anatolyevich Sevostyanov and Alexei Georgievich Kolmakov

Laboratory of Durability and Plasticity of Metal and Composite Materials and Nanomaterials, A.A. Baikov Institute of Metallurgy and Material Science RAS (IMET RAS), Institution of Russian Academy of Sciences, Leninsky Prospect, 49, 119991 Moscow, Russia
* Correspondence: nacakina@mail.ru; Tel.: +7-985-966-5408

Abstract: Composite materials "base–transition layer–surface metal layer (Ta/Ti)" were produced using a complex vacuum technology including magnetron sputtering. The structure (by scanning electron microscopy, Auger electron spectroscopy, X-ray diffractometry) and mechanical properties were studied. An almost linear increase in the thickness of both the surface and transition layers was observed with increasing deposition time and power; however, the growth of the surface layer slowed down with increasing power above some critical value. The transition zone with the growth of time stopped growing upon reaching about 300 nm and was formed approximately 2 times slower than the surface one (and about 3.5 times slower with power). It was noted that with equal sputtering–deposition parameters, the layer growth rates for tantalum and titanium were the same. In the sample with a Ta surface layer deposited on titanium, a strongly textured complex structure with alpha and beta Ta was observed, which is slightly related to the initial substrate structure and the underlying layer. However, even at small thicknesses of the surface layer, the co-deposition of tantalum and titanium contributes to the formation of a single tantalum phase, alpha.

Keywords: tantalum; titanium; magnetron sputtering; surface layer; composite

Citation: Nasakina, E.O.; Sudarchikova, M.A.; Demin, K.Y.; Mikhailova, A.B.; Sergienko, K.V.; Konushkin, S.V.; Kaplan, M.A.; Baikin, A.S.; Sevostyanov, M.A.; Kolmakov, A.G. Study of Co-Deposition of Tantalum and Titanium during the Formation of Layered Composite Materials by Magnetron Sputtering. *Coatings* **2023**, *13*, 114. https://doi.org/10.3390/coatings13010114

Academic Editor: Chenyu Liu

Received: 5 December 2022
Revised: 23 December 2022
Accepted: 27 December 2022
Published: 7 January 2023

Copyright: © 2023 by the authors. Licensee MDPI, Basel, Switzerland. This article is an open access article distributed under the terms and conditions of the Creative Commons Attribution (CC BY) license (https://creativecommons.org/licenses/by/4.0/).

1. Introduction

Tantalum (Ta), titanium (Ti) and their compounds have not ceased to arouse functional interest for humans for decades as a material applicable in various spheres of human life: in optics (transmitting, antireflection, filtering, reflecting, absorbing media) [1–3], electronics (conductors, semiconductors, dielectrics) [4,5], machine and instrument making, construction and everyday life (tribological, wear-resistant, functional, protective coatings, resistant to aggressive environments, decorative, antibacterial, etc.) [6–10], environmental cleanup and agriculture [11–13], medicine (biocompatible, adhesive intermediates) [14–25], and etc., due to its significant characteristics (biocompatibility, resistance to aggressive media, wear resistance, electrical, light and thermal conductivity, photocatalytic activity, radiopacity, strength and/or plasticity, etc.), also including in the form of coatings, thin films and surface layers.

Composites, including layered ones, are unique structures that allow effectively combining, improving and forming in radically new characteristics compared with the original components that are inaccessible to classical materials, which has led to their widespread use [26–43]. In particular, composite structures with surface layers such as TaTiON for optics and electronics, TiTa (at a shape memory ratio), Ta/Ti/TiN/Ti/DLC (diamond-like carbon) and Ta/Ti/DLC for implantology, TaN-(Ta,Ti)N-TiN-Ti for energy, etc. are gaining importance in the modern world [44–47]. For example, to obtain corrosion-resistant biocompatible coatings on superelastic alloys, similar in mechanical properties to living tissues, but containing toxic elements, a mixture of tantalum and titanium is used, because at a certain ratio, this mixture also exhibits similar mechanical characteristics [45]. This mixture is obtained by depositing tantalum layers on a titanium-containing substrate by the magnetron method with intermittent mixing by cycling electron-beam additive technique. The process seems to be quite complicated and requires the preliminary presence of titanium in the substrate. That is, the problem of joint deposition of these two metals (in series or in parallel) is now quite relevant.

A fairly popular method for creating composite surfaces is physical deposition in a vacuum is vacuum ion-plasma methods, especially a variety of magnetron sputtering, which allows, at a fairly low cost of time and resources, effectively obtaining high-quality thin surface layers and coatings of various compositions and structures on a substrate of almost any nature and geometry [48–59]. At the same time, the resulting layers parameters are directly connected with the time and power of sputtering, the deposition distance, the state of the substrate surface and other process parameters, which can vary within wide limits. They determine the phase composition of the new surface. If a multi-component spray system used, the variability of the results increases many times over.

The purpose of this work was to study the features of substance deposition in the region of magnetron sputtering of a tantalum-titanium binary system under varying process conditions and their relationship with the structure of the layers formed during vacuum ion-plasma production of layered metal composite materials.

2. Materials and Methods

In this work, the creation of layered composite materials of various nature was carried out using ion-vacuum technologies by forming surface layers of tantalum and titanium on various substrates (base material) using a DC magnetron in an argon gas environment at a Torr International facility (New Windsor, NY, USA).

The use of magnetron sputtering to create surface layers makes it possible to avoid overheating of the substrate by bombarding electrons due to their retention at the sputtered target, which is extremely important for substrate materials with low melting temperatures or a phase structure sensitive to temperature changes, such as, for example, in superelastic titanium alloys: heat treatment makes it possible to change static properties and cyclic loading under operating conditions with a broad diapason of deformations and is essential for the stabilization of properties, creation and successful application of the product.

The formation of a new surface of a mixed composition on the substrate was carried out in two ways: sequential deposition of a layer from one metal onto a layer from another and simultaneous deposition of both metals. To test the sputtering modes before obtaining composite materials with a mixed surface, surface layers of tantalum or titanium were deposited in the form of separate single layers.

Disks made of chemically pure tantalum, titanium, or a bicomponent structure were used as a sputtered target. As a basis for the composites, plates made of titanium alloys TiNi, TiNbMo, TiNbZr, steel, copper, titanium, etc., 10 mm × 10 mm × 0.5 mm in size, were used. The plates were treated with abrasive sandpaper (grit from 400 to 800) and polished (to a mirror surface) with the addition of diamond suspensions with a particle size of 3, 1 and 0.05 microns to remove flat dents and defects. The depth of surface defects after treatment did not exceed 1 μm. Substrates made of steel, copper and glass/SiO_2 are of interest as a basis for the production of functional materials for a broad diapason of applications (electronics, optics, structural materials, etc.), and superelastic titanium alloys in medicine. For cleaning, activation and polishing, the substrate surface was bombarded with argon ions at U = 900 V and I = 80 mA prior to deposition, i.e., preliminary ion etched.

Surface layers were obtained under the following process conditions (deposition parameters): (1) current I ~ 400–1100 mA, voltage U ~ 360–700 V (power supply power ≈ 135–600 W);

(2) deposition time from 5 to 120 min; (3) deposition distance (distance from the target to the substrate) 40–250 mm. The temperature on the substrate surface did not exceed 150°C. The working and residual pressures in the vacuum chamber were ~0.4 and 4×10^{-4} Pa, respectively.

Morphology, type of destruction during mechanical tests and layer-by-layer elemental composition (including using transverse sections) of the surface of materials were studied using a scanning electron microscope TESCAN VEGA II SBU (TESCAN, Brno, Czech Republic) equipped with an attachment for energy-dispersive analysis INCA Energy (Oxford Instruments, High Wycombe, UK), a JAMP-9500F Auger electron spectrometer (JEOL, Tokyo, Japan) in combination with ion etching under argon bombardment at an angle of 30°, and a GDS 850 A glow-discharge atomic emission spectrometer (Leco, St Joseph, MI, USA) with a high-frequency alternating current source.

X-ray diffraction patterns were obtained on ARL X′TRA (Thermo Fisher Scientific) SARL, Ecublens, Switzerland) and UltimaIV (Rigaku, Tokyo, Japan) instruments, in CuKα radiation in parallel beam geometry. The device was calibrated according to the standard sample NIST SRM-1976a, the error in the position of reflections did not exceed 0.01 °2θ. The crystal lattice parameter was refined by extrapolation to θ = 900 using the Nelson-Riley method in the Origin-2017 program (OriginLab Corporation, Northampton, MA, USA), the magnitude of the crystal lattice microdeformation of the main phase was determined using the Williamson–Hall method in the HighScore Plus program (version 3.0.3, PANanalytical, Almelo, the Netherlands). The quantitative content of crystalline phases was estimated by the method of corundum numbers. Before the study, the surface of the samples was cleaned by washing in ethyl alcohol and distilled water.

Static tests were carried out on a universal testing machine INSTRON 3382 (Instron Corp., Norwood, MA, USA) with a stretching speed of 1 mm/min with an accuracy of the traverse speed of ± 0.2% of the value of the set speed. The processing of test results in determining the characteristics of mechanical properties was carried out in accordance with GOST RF standard 10446-80 (ISO 6892-1:2019(E)) using INSTRON Bluehill 2.0 software. For each experimental point, 3–5 samples were tested. The values of yield strength, tensile strength, relative elongation and Young's modulus were determined.

3. Results and Discussion

In general, identical results were obtained when studying the composition of the obtained surface monolayers: the upper surface layer is enriched with oxygen to a depth of 20 nm due to active surface adsorption; the deeper layer consisted only of the deposited element; between it and the substrate there was a transition layer (containing elements of both the substrate and the deposited substance), which was also enriched with oxygen. Depending on the deposition parameters of several successive layers, regularities were obtained identical to those of the single layer's formation, and transition zones were also observed between the layers of deposited metals (Figure 1). Even when using a titanium alloy substrate, regions of layer-by-layer deposition of titanium (without other alloy elements) and tantalum can be noted on the results of the energy-dispersive analysis of a multilayer composite material (Figure 2).

The generation of the transition layer can be considered as a result of magnetron sputtering, when sputtered particles both condense on the substrate surface, and approach it with some additional energy, and their contact leads to a number of particle interactions [60,61]: "driving in" of sputtered atoms and ions, their "knocking out" (interaction can be elastic and inelastic, with or without energy transfer) and re-deposition or, conversely, the surface particles penetration (both the substrate and previously deposited sputtered elements) into the substrate subsurface structure, the formation of radiation defects stimulated mutual diffusion of the substrate elements and of the deposited layer atoms at their interface, etc. This means that the particles brought into the mobilized state (the deposited substance and the area of the substrate surface), repeatedly colliding and chaotically moving on the substrate surface or near it, are constantly mixed. Ultimately, the surface area is so saturated with the sprayed substance that its interaction with new flows of atoms and ions leads to the formation of a pure surface layer of the composite.

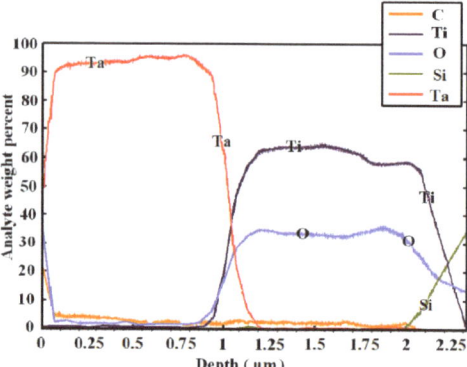

Figure 1. Surface layered chemical composition of the composite material with tantalum and titanium surface layers created for 30 min at 865 mA and 400 V with a distance of 150 mm, on a glass substrate.

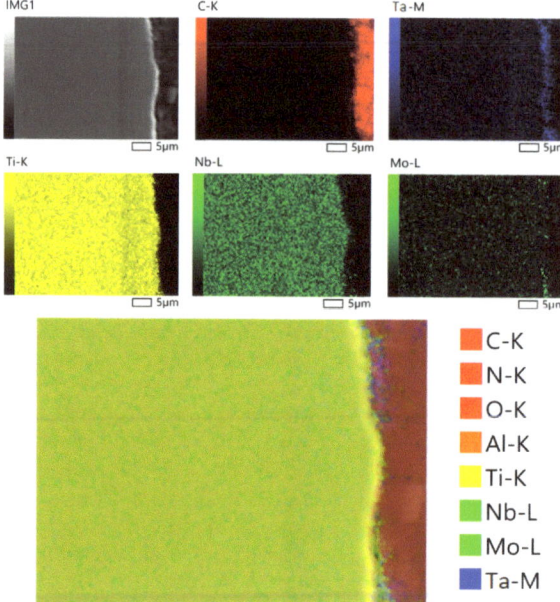

Figure 2. Results of energy-dispersive analysis of a multilayer composite material with tantalum and titanium surface layers created in 30 min at 400 V and 865 mA with a distance of 150 mm, on a TiNbMo base.

The relief of the newly formed surface repeats the surface morphology of the substrate, regardless of the deposition conditions. When applying tantalum to the titanium sublayer, an additional roughness smoothing of the surface is observed (Figure 3). However, at short distances, dotty deepening surface microdefects appear (Figure 4), resembling ion implantation [62], which correlate with a higher flow of spray substance reaching the surface of the substrate compared with longer distances.

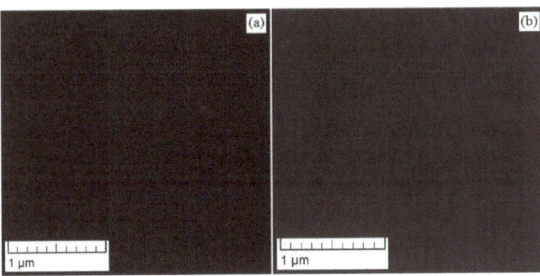

Figure 3. Morphology of a composite material surface with (**a**) a titanium surface layer created in 30 min at 400 V and 865 mA, with a distance of 250 mm, on a glass substrate and (**b**) a tantalum surface layer sputtered on the surface A at the same conditions.

Figure 4. Morphology of the Ta-TiNbZr composite material created in 20 min at a distance of 40 mm.

With an increase in the distance, other conditions being equal, on the one hand, the thickness of the deposited zones diminishes (Figure 5), since more of the sprayed substance diverge to the sides from the main sputtering axis without hitting the substrate; on the other hand, the transition zone thickness rises, which may be due to the higher flux density of the sprayed material particles at a smaller distances, faster and evenly settling on the interface and less penetrating into the base material. The summary layers thickness practically does not change at a distance of 80–150 mm and decreases at a greater distance. As the existence of a significant transition zone is a presumed cause for high adhesion of the newly deposited surface to the base material, and this surface must be adapted to the mechanical behavior of the substrate, and taking into account surface microdefects at short length between the substrate and the sputtered target, distances within 100–150 mm are more optimal.

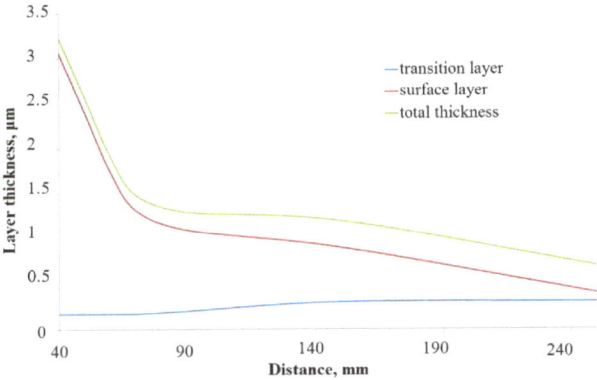

Figure 5. Dependence of the formed layers thickness on distance during magnetron sputtering of tantalum for 30 min at 400 V and 865 mA.

With increasing deposition time an almost linear growth in the thickness of both the surface and transition layers is observed, and the transition zone formed approximately 2 times slower than the surface one. For example (Figure 6), under conditions of 865 mA and 400 V, deposition distance of 200 mm, the growth rate the tantalum surface layer was about 28 nm per minute, and the transition layer was about 15 nm per minute. However, there was a significant difference in that the transition zone stopped growing after about 300 nm, and it can be assumed that the transition layer is saturated. Thus, at a deposition time of 30 minutes, the maximum possible transition layer (0.3 μm) and a surface layer of about 0.9 μm were formed under the given conditions, which correspond to the previously selected optimal conditions when changing the distance. The pattern was preserved under all conditions and materials used.

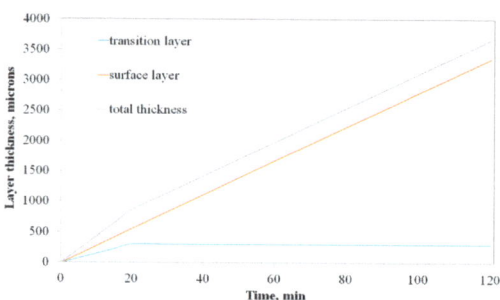

Figure 6. Dependence of layer thickness on time for a composite obtained at 400 V, 865 mA at a distance of 200 mm.

At the beginning of the formation of the new surface zone, the particles of the deposited substance, overcoming the spraying distance, colliding with working gas atoms and ions, with each other and with the new surface of the substrate, do not end up in each surface section at the same time and at first interact with it randomly and irregularly Thereafter (with an increase in the spraying time, and hence the time of exposure to the surface), the particles kept colliding and mixing, trying to take an energetically more favorable state and position, leading to a more uniform distribution of the precipitated substance on the surface. The selected value of the operating pressure for the surface layer deposition of ≈ 0.4 Pa, according to the literature data, promotes the formation of strong films of a crystalline structure with low surface roughness and high density [55,58,59,63]. The continuous interaction of the mobilized particles of the target and the substrate contributes to the fact that, when the layer thickness reached 300 nm in this work, the islands were already smoothed out.

The total thickness value of the surface and transition layers increased almost linearly with increasing deposition power. Thus, when titanium was deposited on any substrate at the deposition distance 150 mm, for 30 min, in the deposition power range of 0–350 W, the average increase in the thickness of the surface layer was 2.61 nm/W, and the transition layer was 0.725 nm/W (Figure 7). An increase in their thickness may be associated with an increased target sputtering rate. The effect of a further increase in power (up to 500 W) was slightly less, which may be caused by the compaction of the near-surface zone and the reduction in the time for transition layer creation with an increase in the deposited material flux density and energy, but at the same time, the target consumption of the target and the possibility of contamination of the composite surface increased, including by-elements of the stainless steel vacuum chamber walls, which can be etched by high-energy particles; it is also possible to spray the newly formed surface with high-energy particles of the deposited flow.

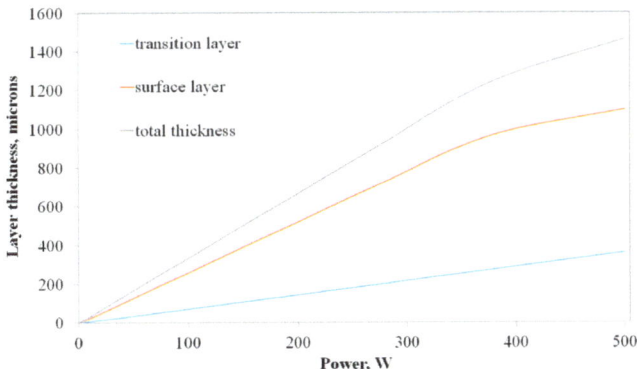

Figure 7. Dependence of layer thickness on power for a composite obtained for 30 min of tantalum sputtering at a distance of 150 mm.

It was noted that with equal sputtering–deposition parameters, the layer growth rates for tantalum and titanium are the same. For example, this is shown in Figure 1, where tantalum and titanium layers obtained with the same parameters are identical in thickness, which is in good agreement with the literature data, for example, with ref. [44].

With the simultaneous deposition of tantalum and titanium, composite materials "oxy nitride layer (the area at the very boundary of the solid body with the surrounding gaseous medium, free from substrate elements, where the content of titanium or tantalum is not at a maximum, about 10 nm thick)—a surface layer of tantalum with titanium–a transition layer containing elements of the surface and the base–the base" were obtained (Figures 8–10). The general regularity of the change in the in composition with depth of the obtained composites is approximately the same. Depending on the deposition parameters, regularities were obtained identical to the regularities of the formation of single layers of individual elements.

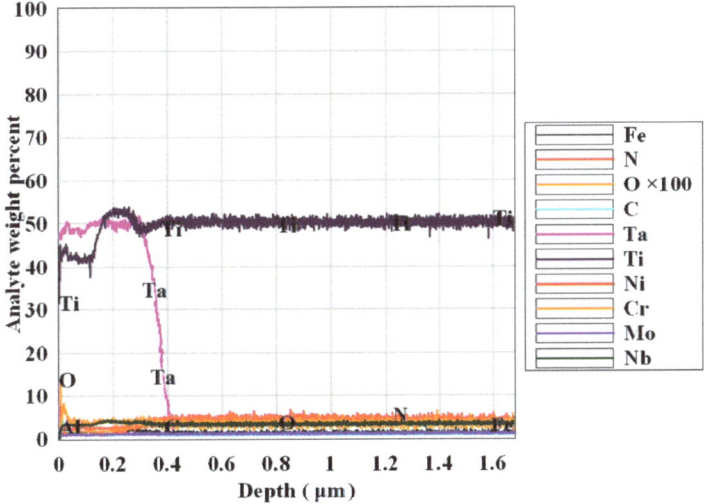

Figure 8. Layered composition of the surface of a composite material with a tantalum–titanium surface layer created in 30 min at 400 V and 865 mA with a distance of 250 mm, on a TiNbMo substrate.

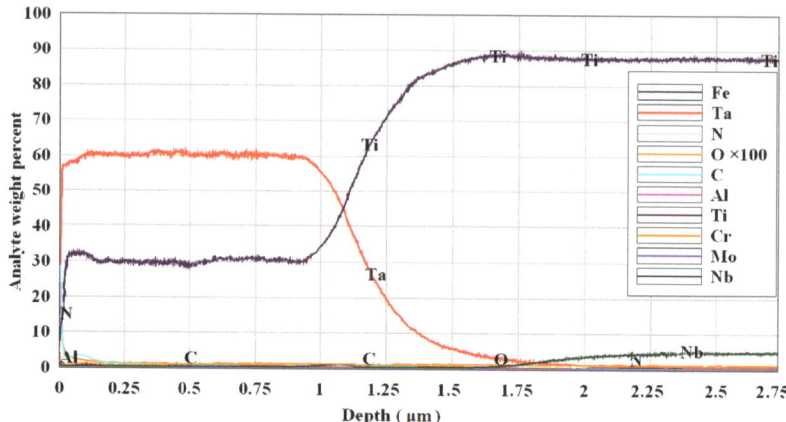

Figure 9. Layered composition of the surface of a composite material with a tantalum–titanium surface layer created in 30 min at 400 V and 865 mA with a distance of 150 mm, on a TiNbMo substrate.

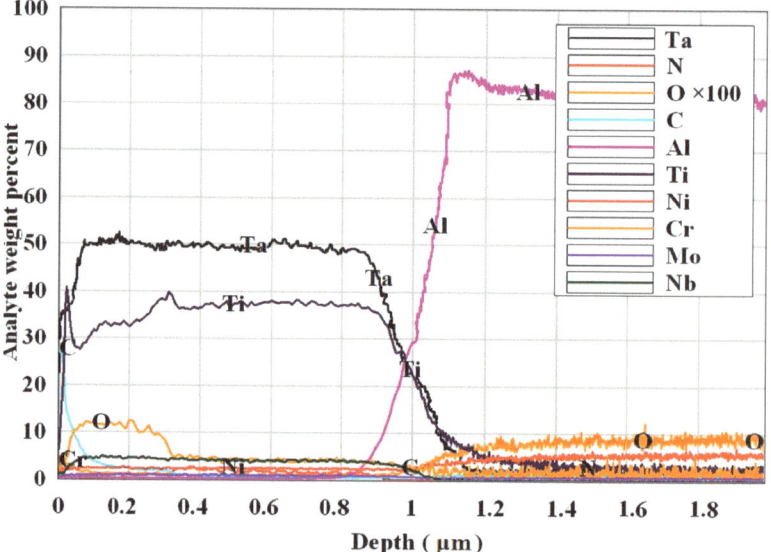

Figure 10. Layered composition of the surface of a composite material with a tantalum–titanium surface layer, created in 30 min at 400 V and 865 mA with a distance of 150 mm, on an aluminum substrate.

The results of element mapping in a transverse section (Figures 11–13) are in good agreement with the layer-by-layer analysis of the chemical composition of the surface. The surface layer of the mixed composition clearly stands out on the substrate of both foreign material (aluminum) and titanium alloy. When mapping surface elements, a uniform distribution of deposited metals without the formation of clusters can be noted, and a visual decrease in the contribution of the substrate can be seen when the layer thickness increased from 0.4 to 0.9 µm (Figures 14–16).

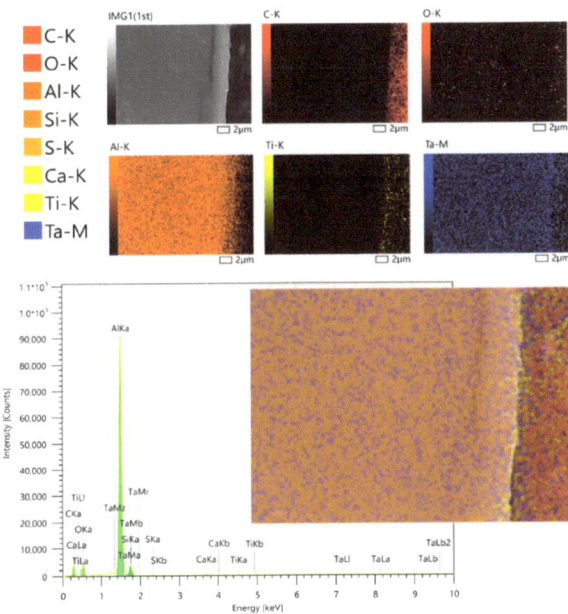

Figure 11. Distribution of elements in a cross section of a composite material with a tantalum–titanium surface layer, created in 30 min at 400 V and 865 mA with a distance of 250 mm (the thickness is 0.4 μm), on an aluminum substrate.

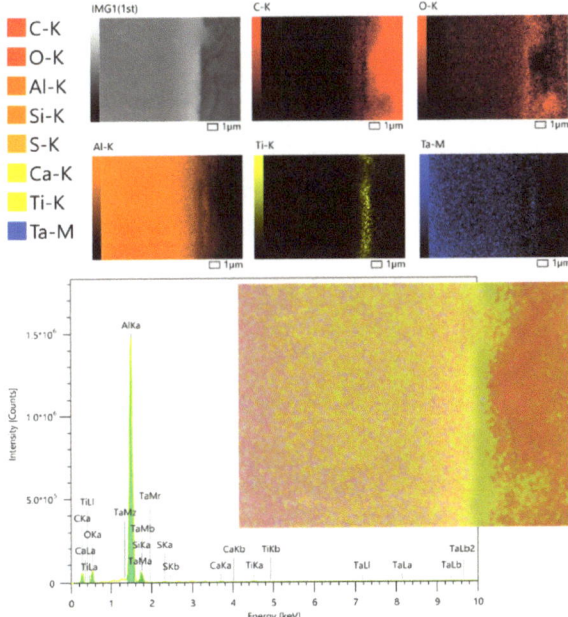

Figure 12. Distribution of elements in a cross section of a composite material with a tantalum–titanium surface layer, created in 30 min at 400 V and 865 mA with a distance of 150 mm (the thickness is 0.9 μm), on an aluminum substrate.

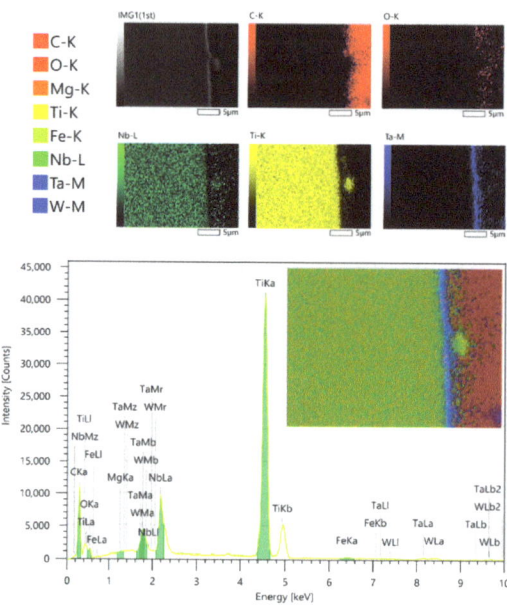

Figure 13. Distribution of elements in a cross section of a composite material with a tantalum–titanium surface layer created in 30 min at 400 V and 865 mA with a distance of 150 mm (the thickness is 0.9 µm), on a TiNbMo substrate.

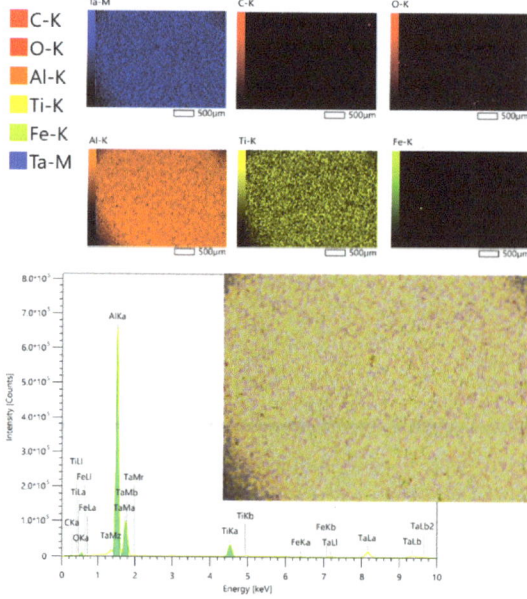

Figure 14. Distribution of elements in the surface of a composite material with a tantalum–titanium surface layer, created in 30 min at 400 V and 865 mA with a distance of 250 mm (the thickness is 0.4 µm), on an aluminum substrate.

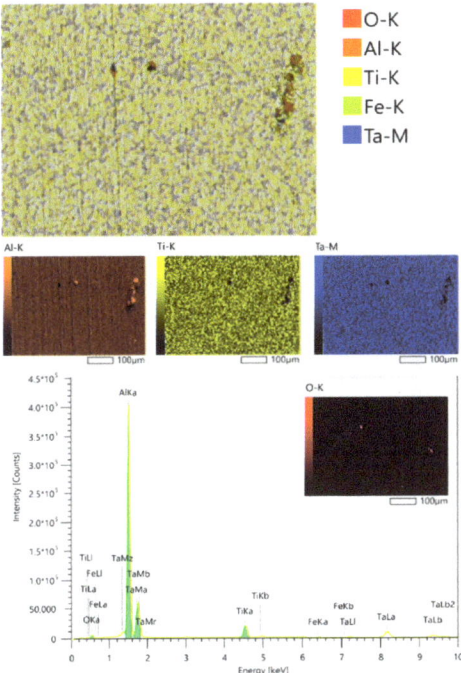

Figure 15. Distribution of elements in the surface of a composite material with a tantalum–titanium surface layer, created in 30 min at 400 V and 865 mA with a distance of 150 mm (the thickness is 0.9 µm), on an aluminum substrate.

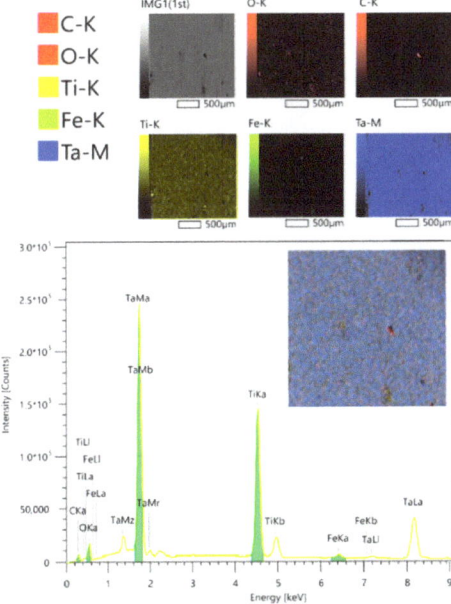

Figure 16. Distribution of elements in the surface of a composite material with a tantalum–titanium surface layer, created in 30 min at 400 V and 865 mA with a distance of 150 mm (the thickness is 0.9 µm), on a TiNbMo substrate.

In the case of X-ray phase analysis of a titanium surface layer, regardless of its thickness, production parameters and the nature of the underlying substrate, only the beta phase (cubic crystal lattice) is observed, and this composition was not identical to the cast sputtered target phase composition (alpha). An example is shown in the Figure 17, the main peaks correspond to the phases: Ti—Im-3m: 39°26′, 83°31′; B-19 NiTi—21/m(11): 21°46′, 36°18′, 39°26′, 40°45′, 43°8′, 44°17′, 45°8′, 46°51′, 56°43′, 83°31′; R NiTi—P-3(147): 39°26′, 44°34′, 45°8′, 46°51′, 78°9′, 79°46′, 93°36′.

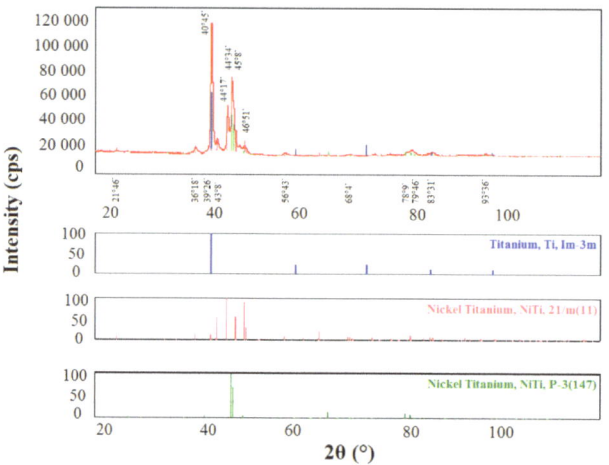

Figure 17. Sample with TiNi base and Ti surface layer created in 30 min, at 400 V and 865 mA with a distance of 150 mm.

When creating thin tantalum films and layers, as noted in the literature, its formation in both beta and alpha states is possible [51–59,63–73]. At the same time, several theories have been developed for the formation of tantalum in one or another phase state, mainly related to temperature and pressure (which determine the motility and energy of atoms) and the substrate composition and surface condition. However, different authors often come to conflicting results.

According to our previous studies [74,75], regardless of the deposition conditions, the beta phase of tantalum is first formed, and alpha-tantalum is deposited on it (which is presumably due to the oxygen presence in the surface of all substrates), as in ref. [56], where, however, this was associated with a significant heating of the surface (more than 350 °C).

It was observed that tantalum in alpha state is created at temperature above 400 °C, which contributes to an increase in the mobility of the deposited atoms: either during initially heating of the substrate or annealing following deposition (when the obtained beta phase turns into alpha Ta) [56,63,67]. However, at the temperature range from 400 to 500 °C the beta tantalum was also obtained [63,67], while α was formed even without heating [54,57]. It was indicated that with increasing temperature, the grain size, the amount of surface layer impurities, and its amorphism decrease.

A high oxygen content in the working atmosphere in [58] leads to the rapid creation of oxides causing the creation of a beta tantalum layer, while in ref. [57] the oxygen environment did not interfere with the creation tantalum alpha phase. When deposited on glass and silicon substrates, a pressure of 0.5 ± 0.7 Pa in Refs [55,58,59] led to the formation of α-Ta and at lower or higher pressure, β-Ta; however, in ref. [57], the α phase was already formed at 0.28 Pa, and in [59], the alpha tantalum creation also occurred at pressures of 0.3 and 1.4 Pa.

Being the zone of nucleation of a new surface, the base substrate surface determines the new structure creation nature. It has been noted that beta tantalum is formed on amorphous surfaces containing carbon or oxygen (the inartificial state of glass or titanium and aluminum in an oxygen atmosphere), whereas, for example, on titanium without natural oxide, on previously deposited α-Ta or TaN alpha tantalum is formed [52,54,56,59]. Additionally, it was pointed out that (110) phase is the lowest energy lattice for bcc materials and causes the formation of the same structure on itself, and α-Ta (110) is the thermodynamically most stable phase.

In an oxygen-free environment, beta and alpha titanium formed, respectively, and on glass and silicon, α-Ta formed [54]. Beta titanium and alpha tantalum have a similar type of crystal lattice (110); the α-Ti lattice parameters match with the parameters of the hexagonal lattice composed of atoms of the nearest α-Ta planes. In these two cases, titanium grains can serve as the nucleation core for Ta crystallites. The amorphous oxide layer differs too much in structure from the α-Ta crystal lattice and this difference leads to the formation of β-Ta.

Thus, in this work, we expected to see a modification of tantalum.

In this work, if a Ta surface layer deposited on titanium, after prolonged ion etching (i.e. without an oxygen-containing surface) and without it, a lot of alpha and beta tantalum peaks are recorded, which corresponds to different crystal orientations: beta (002, 410, 202, 004, 513, 333, 404, etc.), alpha (110, 211, and 220), i.e., a strongly textured structure is formed, regardless of the substrate surface. Example is shown in Figure 18, the main peaks correspond to the phases: Ti-beta, Im-3m: 38°28′, 69°55′, 121°22′; Ta-alpha, Im-3m: 38°28′, 69°55′, 121°22′; Ta- beta, hP2/1: 33°42′, 69°55′, 121°22′; Ta-beta, tP30/17: 33°42′, 38°28′, 69°55′, 121°22′).

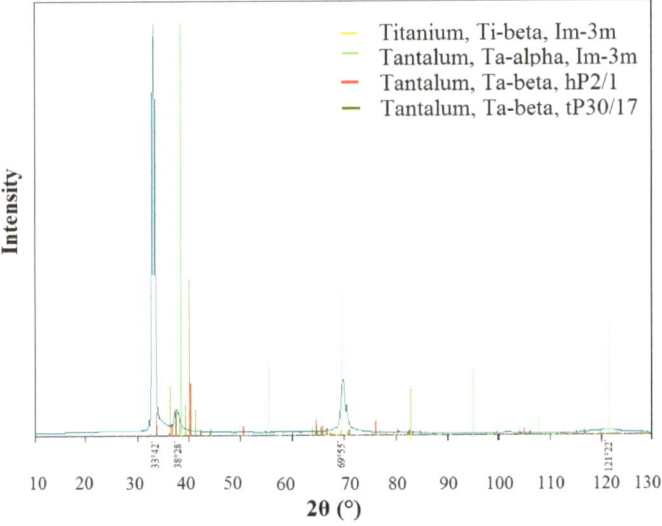

Figure 18. X-ray patterns of samples of a multilayer composite material with tantalum on titanium surface layers created in 30 min, at 400 V and 865 mA with a distance of 250 mm, on a glass substrate.

X-ray phase studies suggest that even at small thicknesses of the surface layer, the co-deposition of tantalum and titanium contributes to the formation of a single tantalum phase, alpha. In Figure 19 the main peak on the X-ray corresponds to the substrate due to averaging over the entire depth of analysis, corresponding to a mixture of alpha titanium (hexagonal lattice) and beta, the surface peaks correspond to the TiTa (cubic), beta Ti (cubic, typical for magnetron deposition according to previous studies) and alpha Ta (cubic) (i.e., the main angles in the figure correspond to Ti-alpha, P63-mmc: 35°6′, 38°32′, 40°8′, 53°, 63°40′, 70°32′, 76°34′, 82°36′, Ta-alpha, Im-3m: 38°32′, 56°24′, 70°32′, 82°36′, 96°34′, TiTa, Im-3m: 35°6′, 36°16′, 38°32′, 56°24′, 70°32′, 82°36′, 96°34′, Ti-beta, Im-3m: 38°32′, 56°24′, 70°32′, 82°36′, 96°34′), while, as previously discussed, beta-Ti and alpha-Ta have a similar type of crystal lattice (110), and the parameters of the alpha-Ti lattice coincide with the parameters of the hexagonal lattice composed of atoms of the nearest alpha-Ta planes, which determine the only possible option for the development of tantalum crystallite. If the deposition occurs on a substrate of a different nature (Figure 20, on aluminum), we also observe the main peak from the substrate, and, on the surface, a mixture of alpha tantalum, beta titanium and TiTa (the main angles correspond to Al, Fm-3m: 38°28′, 44°38′, 65°12′, 78°, TiTa, Im-3m: 38°28′, 56°6′, 70°, 82°34′, 96°19′, Ta-alpha, Im-3m: 38°28′, 56°6′, 70°, 82°34′, 96°19′, and Ti-beta, Im-3m: 38°28′, 56°6′, 70°, 82°34′, 96°19′).

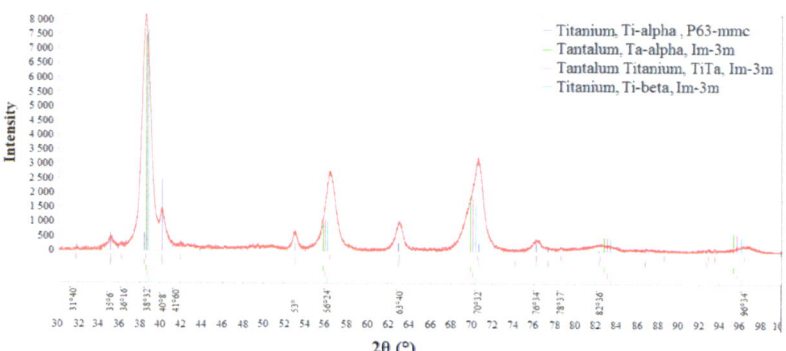

Figure 19. TiNbZr sample after magnetron sputtering of a Ta–Ti target created in 30 min, at 400 V and 865 mA with a distance of 250 mm.

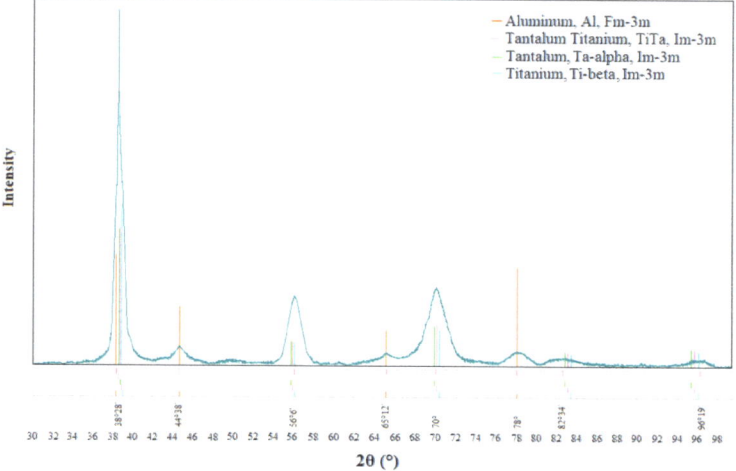

Figure 20. Al sample after magnetron sputtering of a Ta–Ti target created in 30 min, at 400 V and 865 mA with a distance of 250 mm.

Carrying out irradiation of the substrate surface with argon ions before layer deposition not only to cleaning the surface from impurities, but also to fine polishing and activation of this surface, which ensures the formation of a stable transition layer.

The results of studying the mechanical properties of composites with titanium and tantalum–titanium surface layers and a titanium alloy base are shown in Table 1.

Table 1. Mechanical properties of Ti-Nb-Zr wire after surface treatment and composite material based on it.

No.	Sample	Rel. ext.	Yield Strength (MPa)	Tensile Strength (MPa)	Load (kgf)	Young's Modulus (GPa)
1	Ti-Nb-Zr	3.05	583.68	671.58	44.58	27.571
2	Ti-Nb-Zr–Ti, 30 min	3.21	419.79	579.43	47.19	25.675
3	Ti-Nb-Zr–Ti–Ta, 30 min	3.65	448.75	619.67	45.42	23.278
4	Ti-Nb-Zr–Ta/Ti, 30 min	3.67	451.24	618.72	47.12	22.589

Based on the data obtained, it can be concluded that the elongation increases as the surface layers are deposited and thickened by the magnetron deposition method, while the strength and Young's modulus decrease slightly, and the additional deposition of tantalum increases all indicators.

4. Conclusions

The patterns of formation of layered composite materials with a surface metal layer of a bicomponent composition (Ta/Ti) using a complex vacuum technology, including magnetron sputtering, with varying process conditions, including options for supplying elemental fluxes (in series and in parallel, from one source when using a target of mixed composition) were studied.

An almost linear increase in the thickness of both the surface and transition layers was observed with increasing deposition time, and the transition zone was formed approximately two times slower than the surface one and stopped growing after reaching about 300 nm. The total thickness of the surface and transition layers increased almost linearly with increasing power, but after some power critical value the growth rate decreased. With equal sputtering–deposition parameters, the layer growth rates for tantalum and titanium were the same.

In the case of a sample with a Ta surface layer deposited on titanium, a strongly textured mixture of alpha and beta phases was observed, which was insignificantly related to the initial structure of the substrate and the underlying layer. However, even at small thicknesses of the surface layer, the joint deposition of tantalum and titanium contributed to the formation of a single tantalum phase, alpha.

The elongation increased as the surface layers were deposited and thickened, while the strength and Young's modulus decreased slightly, and the additional deposition of tantalum on titanium improved everything.

The results of the works carried out using vacuum technologies and modern methods for studying materials have prospects for use in various fields of science and technology (medicine, electronics, optics, structural materials, etc.).

Author Contributions: Conceptualization, M.A.S. (Maria Andreevna Sudarchikova); funding acquisition, E.O.N.; investigation, E.O.N., K.V.S., K.Y.D., A.B.M., M.A.K., A.S.B. and S.V.K.; methodology, A.G.K., E.O.N.; project administration, M.A.S. (Mikhail Anatolyevich Sevostyanov); supervision, M.A.S. (Mikhail Anatolyevich Sevostyanov); validation, S.V.K.; visualization, K.V.S.; writing—original draft, E.O.N.; writing—review and editing, E.O.N. and M.A.S. (Maria Andreevna Sudarchikova). All authors have read and agreed to the published version of the manuscript.

Funding: This research was funded by Russian Science Foundation, grant number 21-79-10256.

Institutional Review Board Statement: Not applicable.

Informed Consent Statement: Not applicable.

Data Availability Statement: Not applicable.

Conflicts of Interest: The authors declare no conflict of interest.

References

1. Wiatrowski, A.; Mazur, M.; Obstarczyk, A.; Wojcieszak, D.; Kaczmarek, D.; Morgiel, J.; Gibson, D. Comparison of the Physicochemical Properties of TiO2 Thin Films Obtained by Magnetron Sputtering with Continuous and Pulsed Gas Flow. *Coatings* **2018**, *8*, 412. [CrossRef]
2. El-Fattah, H.A.A.; El-Mahallawi, I.S.; Shazly, M.H.; Khalifa, W.A. Optical Properties and Microstructure of TiNxOy and TiN Thin Films before and after Annealing at Different Conditions. *Coatings* **2019**, *9*, 22. [CrossRef]
3. Hu, Y.; Rasadujjaman, M.; Wang, Y.; Zhang, J.; Yan, J.; Baklanov, M. Study on the Electrical, Structural, Chemical and Optical Properties of PVD Ta(N) Films Deposited with Different N_2 Flow Rates. *Coatings* **2021**, *11*, 937. [CrossRef]
4. Dongquoc, V.; Seo, D.-B.; Anh, C.V.; Lee, J.-H.; Park, J.-H.; Kim, E.-T. Controlled Surface Morphology and Electrical Properties of Sputtered Titanium Nitride Thin Film for Metal–Insulator–Metal Structures. *Appl. Sci.* **2022**, *12*, 10415. [CrossRef]
5. Korotkova, K.; Bainov, D.; Smirnov, S.; Yunusov, I.; Zhidik, Y. Electrical Conductivity and Optical Properties of Nanoscale Titanium Films on Sapphire for Localized Plasmon Resonance-Based Sensors. *Coatings* **2020**, *10*, 1165. [CrossRef]
6. Zaman, A.; Meletis, E.I. Microstructure and Mechanical Properties of TaN Thin Films Prepared by Reactive Magnetron Sputtering. *Coatings* **2017**, *7*, 209. [CrossRef]
7. Qiao, Z.; Li, X.; Lv, Y.; Xie, Y.; Hu, Y.; Wang, J.; Li, H.; Wang, S. Depositing a Titanium Coating on the Lithium Neutron Production Target by Magnetron Sputtering Technology. *Materials* **2021**, *14*, 1873. [CrossRef] [PubMed]
8. Kaltschmidt, B.P.; Asghari, E.; Kiel, A.; Cremer, J.; Anselmetti, D.; Kaltschmidt, C.; Kaltschmidt, B.; Hütten, A. Magnetron Sputtering of Transition Metals as an Alternative Production Means for Antibacterial Surfaces. *Microorganisms* **2022**, *10*, 1843. [CrossRef]
9. de Monteynard, A.; Luo, H.; Chehimi, M.; Ghanbaja, J.; Achache, S.; François, M.; Billard, A.; Sanchette, F. The Structure, Morphology, and Mechanical Properties of Ta-Hf-C Coatings Deposited by Pulsed Direct Current Reactive Magnetron Sputtering. *Coatings* **2020**, *10*, 212. [CrossRef]
10. Rivera-Tello, C.D.; Broitman, E.; Flores-Ruiz, F.J.; Perez-Alvarez, J.; Flores-Jiménez, M.; Jiménez, O.; Flores, M. Micro and Macro-Tribology Behavior of a Hierarchical Architecture of a Multilayer TaN/Ta Hard Coating. *Coatings* **2020**, *10*, 263. [CrossRef]
11. Farahani, N.; Kelly, P.J.; West, G.; Hill, C.; Vishnyakov, V. Photocatalytic Activity of Reactively Sputtered Titania Coatings Deposited Using a Full Face Erosion Magnetron. *Coatings* **2013**, *3*, 177–193. [CrossRef]
12. Wang, Y.-H.; Rahman, K.H.; Wu, C.-C.; Chen, K.-C. A Review on the Pathways of the Improved Structural Characteristics and Photocatalytic Performance of Titanium Dioxide (TiO2) Thin Films Fabricated by the Magnetron-Sputtering Technique. *Catalysts* **2020**, *10*, 598. [CrossRef]
13. Kelly, P.J.; West, G.T.; Ratova, M.; Fisher, L.; Ostovarpour, S.; Verran, J. Structural Formation and Photocatalytic Activity of Magnetron Sputtered Titania and Doped-Titania Coatings. *Molecules* **2014**, *19*, 16327–16348. [CrossRef] [PubMed]
14. Kim, J.-Y.; Park, J.-B. Various Coated Barrier Membranes for Better Guided Bone Regeneration: A Review. *Coatings* **2022**, *12*, 1059. [CrossRef]
15. García, E.; Flores, M.; Rodríguez, E.; Rivera, L.P.; Camps, E.; Muhl, S. Tribological, Tribocorrosion and Wear Mechanism Studies of TaZrN Coatings Deposited by Magnetron Sputtering on TiAlV Alloy. *Coatings* **2018**, *8*, 295. [CrossRef]
16. Alvarez, R.; Muñoz-Piña, S.; González, M.U.; Izquierdo-Barba, I.; Fernández-Martínez, I.; Rico, V.; Arcos, D.; García-Valenzuela, A.; Palmero, A.; Vallet-Regi, M.; et al. Antibacterial Nanostructured Ti Coatings by Magnetron Sputtering: From Laboratory Scales to Industrial Reactors. *Nanomaterials* **2019**, *9*, 1217. [CrossRef]
17. Yelkarasi, C.; Recek, N.; Kazmanli, K.; Kovač, J.; Mozetič, M.; Urgen, M.; Junkar, I. Biocompatibility and Mechanical Stability of Nanopatterned Titanium Films on Stainless Steel Vascular Stents. *Int. J. Mol. Sci.* **2022**, *23*, 4595. [CrossRef] [PubMed]
18. Mina-Aponzá, S.; Castro-Narváez, S.; Caicedo-Bejarano, L.; Bermeo-Acosta, F. Study of Titanium–Silver Monolayer and Multilayer Films for Protective Applications in Biomedical Devices. *Molecules* **2021**, *26*, 4813. [CrossRef] [PubMed]
19. Ding, Z.; Zhou, Q.; Wang, Y.; Ding, Z.; Tang, Y.; He, Q. Microstructure and properties of monolayer, bilayer and multilayer Ta2O5-based coatings on biomedical Ti-6Al-4V alloy by magnetron sputtering. *Ceram. Int.* **2020**, *47*, 1133–1144. [CrossRef]
20. Zhang, M.; Ma, Y.; Gao, J.; Hei, H.; Jia, W.; Bai, J.; Liu, Z.; Huang, X.; Xue, Y.; Yu, S.; et al. Mechanical, Electrochemical, and Osteoblastic Properties of Gradient Tantalum Coatings on Ti6Al4V Prepared by Plasma Alloying Technique. *Coatings* **2021**, *11*, 631. [CrossRef]
21. Rodrigues, M.M.; Fontoura, C.P.; Maddalozzo, A.E.D.; Leidens, L.M.; Quevedo, H.G.; Souza, K.D.S.; Crespo, J.D.S.; Michels, A.F.; Figueroa, C.A.; Aguzzoli, C. Ti, Zr and Ta coated UHMWPE aiming surface improvement for biomedical purposes. *Compos. Part B: Eng.* **2020**, *189*, 107909. [CrossRef]
22. Ji, P.; Liu, S.; Deng, H.; Ren, H.; Zhang, J.; Sun, T.; Xu, K.; Shi, C. Effect of magnetron-sputtered monolayer Ta and multilayer Ti-Zr-Ta and Zr-Ti-Ta coatings on the surface properties of biomedical Ti-6Al-4V alloy. *Mater. Lett.* **2022**, *322*. [CrossRef]
23. Baigonakova, G.; Marchenko, E.; Yasenchuk, Y.; Kokorev, O.; Vorozhtsov, A.; Kulbakin, D. Microstructural characterization, wettability and cytocompatibility of gradient coatings synthesized by gas nitriding of three-layer Ti/Ni/Ti nanolaminates magnetron sputtered on the TiNi substrate. *Surf. Coatings Technol.* **2022**, *436*. [CrossRef]
24. Lenis, J.; Rico, P.; Ribelles, J.G.; Pacha-Olivenza, M.; González-Martín, M.; Bolívar, F. Structure, morphology, adhesion and in vitro biological evaluation of antibacterial multi-layer HA-Ag/SiO2/TiN/Ti coatings obtained by RF magnetron sputtering for biomedical applications. *Mater. Sci. Eng. C* **2020**, *116*, 111268. [CrossRef] [PubMed]

25. Lenis, J.; Bejarano, G.; Rico, P.; Ribelles, J.G.; Bolívar, F. Development of multilayer Hydroxyapatite - Ag/TiN-Ti coatings deposited by radio frequency magnetron sputtering with potential application in the biomedical field. *Surf. Coatings Technol.* **2019**, *377*, 124856. [CrossRef]
26. López-Huerta, F.; Cervantes, B.; González, O.; Hernández-Torres, J.; García-González, L.; Vega, R.; Herrera-May, A.L.; Soto, E. Biocompatibility and Surface Properties of TiO2 Thin Films Deposited by DC Magnetron Sputtering. *Materials* **2014**, *7*, 4105–4117. [CrossRef] [PubMed]
27. Marchenko, E.; Baigonakova, G.; Kokorev, O.; Yasenchuk, Y.; Vorozhtsov, A. Biocompatibility Assessment of Coatings Obtained in Argon and Nitrogen Atmospheres for TiNi Materials. *Metals* **2022**, *12*, 1603. [CrossRef]
28. Domínguez-Crespo, M.; Torres-Huerta, A.; Rodríguez, E.; González-Hernández, A.; Brachetti-Sibaja, S.; Dorantes-Rosales, H.; López-Oyama, A. Effect of deposition parameters on structural, mechanical and electrochemical properties in Ti/TiN thin films on AISI 316L substrates produced by r. f. magnetron sputtering. *J. Alloy. Compd.* **2018**, *746*, 688–698. [CrossRef]
29. Akimchenko, I.O.; Rutkowski, S.; Tran, T.-H.; Dubinenko, G.E.; Petrov, V.I.; Kozelskaya, A.I.; Tverdokhlebov, S.I. Polyether Ether Ketone Coated with Ultra-Thin Films of Titanium Oxide and Zirconium Oxide Fabricated by DC Magnetron Sputtering for Biomedical Application. *Materials* **2022**, *15*, 8029. [CrossRef] [PubMed]
30. Subramanian, B.; Muraleedharan, C.; Ananthakumar, R.; Jayachandran, M. A comparative study of titanium nitride (TiN), titanium oxy nitride (TiON) and titanium aluminum nitride (TiAlN), as surface coatings for bio implants. *Surf. Coatings Technol.* **2011**, *205*, 5014–5020. [CrossRef]
31. González-Hernández, A.; Aperador, W.; Flores, M.; Onofre-Bustamante, E.; Bermea, J.E.; Bautista-García, R.; Gamboa-Soto, F. Influence of Deposition Parameters on Structural and Electrochemical Properties of Ti/Ti$_2$N Films Deposited by RF-Magnetron Sputtering. *Metals* **2022**, *12*, 1237. [CrossRef]
32. Stolin, A.M.; Bazhin, P. Manufacture of multipurpose composite and ceramic materials in the combustion regime and high-temperature deformation (SHS extrusion). *Theor. Found. Chem. Eng.* **2014**, *48*, 751–763. [CrossRef]
33. Bazhin, P.; Stolin, A.M.; Alymov, M.I. Preparation of nanostructured composite ceramic materials and products under conditions of a combination of combustion and high-temperature deformation (SHS extrusion). *Nanotechnologies Russ.* **2014**, *9*, 583–600. [CrossRef]
34. Krivoshapkin, P.; Mikhaylov, V.; Krivoshapkina, E.; Zaikovskii, V.; Melgunov, M.; Stalugin, V. Mesoporous Fe–alumina films prepared via sol–gel route. *Microporous Mesoporous Mater.* **2015**, *204*, 276–281. [CrossRef]
35. Kononova, S.V.; Korytkova, E.N.; Maslennikova, T.; Romashkova, K.A.; Kruchinina, E.V.; Potokin, I.L.; Gusarov, V. Polymer-inorganic nanocomposites based on aromatic polyamidoimides effective in the processes of liquids separation. *Russ. J. Gen. Chem.* **2010**, *80*, 1136–1142. [CrossRef]
36. Fattah-Alhosseini, A.; Elmkhah, H.; Ansari, G.; Attarzadeh, F.; Imantalab, O. A comparison of electrochemical behavior of coated nanostructured Ta on Ti substrate with pure uncoated Ta in Ringer's physiological solution. *J. Alloy. Compd.* **2018**, *739*, 918–925. [CrossRef]
37. Seidl, W.; Bartosik, M.; Kolozsvári, S.; Bolvardi, H.; Mayrhofer, P. Improved mechanical properties, thermal stabilities, and oxidation resistance of arc evaporated Ti-Al-N coatings through alloying with Ta. *Surf. Coatings Technol.* **2018**, *344*, 244–249. [CrossRef]
38. Sevost'Yanov, M.A.; Nasakina, E.O.; Baikin, A.S.; Sergienko, K.V.; Konushkin, S.V.; Kaplan, M.A.; Seregin, A.V.; Leonov, A.V.; Kozlov, V.A.; Shkirin, A.V.; et al. Biocompatibility of new materials based on nano-structured nitinol with titanium and tantalum composite surface layers: Experimental analysis in vitro and in vivo. *J. Mater. Sci. Mater. Med.* **2018**, *29*, 33. [CrossRef]
39. Li, P.; Zhang, X.; Xu, R.; Wang, W.; Liu, X.; Yeung, K.W.K.; Chu, P.K. Electrochemically deposited chitosan/Ag complex coatings on biomedical NiTi alloy for antibacterial application. *Surf. Coat. Technol.* **2013**, *232*, 370–375. [CrossRef]
40. Cheng, Y.; Cai, W.; Li, H.T.; Zheng, Y.F. Surface modification of NiTi alloy with tantalum to improve its biocompatibility and radiopacity. *J. Mater. Sci.* **2006**, *41*, 4961–4964. [CrossRef]
41. Lee, D.-W.; Kim, Y.-N.; Cho, M.-Y.; Ko, P.-J.; Lee, D.; Koo, S.-M.; Moon, K.-S.; Oh, J.-M. Reliability and characteristics of magnetron sputter deposited tantalum nitride for thin film resistors. *Thin Solid Films* **2018**, *660*, 688–694. [CrossRef]
42. Siddiqui, J.; Hussain, T.; Ahmad, R.; Umar, Z.A. On the structural, morphological and electrical properties of tantalum oxy nitride thin films by varying oxygen percentage in reactive gases plasma. *Chin. J. Phys.* **2017**, *55*, 1412–1422. [CrossRef]
43. Kumar, M.; Kumari, N.; Kumar, V.P.; Karar, V.; Sharma, A.L. Determination of optical constants of tantalum oxide thin film deposited by electron beam evaporation. *Mater. Today: Proc.* **2018**, *5*, 3764–3769. [CrossRef]
44. Cristea, D.; Velicu, I.-L.; Cunha, L.; Barradas, N.; Alves, E.; Craciun, V. Tantalum-Titanium Oxynitride Thin Films Deposited by DC Reactive Magnetron Co-Sputtering: Mechanical, Optical, and Electrical Characterization. *Coatings* **2021**, *12*, 36. [CrossRef]
45. Ormanova, M.; Dechev, D.; Ivanov, N.; Mihai, G.; Gospodinov, M.; Valkov, S.; Enachescu, M. Synthesis and Characterization of Ti-Ta-Shape Memory Surface Alloys Formed by the Electron-Beam Additive Technique. *Coatings* **2022**, *12*, 678. [CrossRef]
46. Tu, R.; Min, R.; Yang, M.; Yuan, Y.; Zheng, L.; Li, Q.; Ji, B.; Zhang, S.; Yang, M.; Shi, J. Overcoming the Dilemma between Low Electrical Resistance and High Corrosion Resistance Using a Ta/(Ta,Ti)N/TiN/Ti Multilayer for Proton Exchange Membrane Fuel Cells. *Coatings* **2022**, *12*, 689. [CrossRef]
47. Guo, D.; Zhang, S.; Huang, T.; Wu, S.; Ma, X.; Guo, F. Corrosion Properties of DLC Film in Weak Acid and Alkali Solutions. *Coatings* **2022**, *12*, 1776. [CrossRef]
48. Bunshah, R.F. *Deposition Technologies for Films and Coating*; Noyes Publikations: Park Ridge, IL, USA, 1982; p. 489 p.

49. Kulczyk-Malecka, J.; Kelly, P.J.; West, G.; Clarke, G.C.; Ridealgh, J.A. Characterisation Studies of the Structure and Properties of As-Deposited and Annealed Pulsed Magnetron Sputtered Titania Coatings. *Coatings* **2013**, *3*, 166–176. [CrossRef]
50. Al-Masha'Al, A.; Bunting, A.; Cheung, R. Evaluation of residual stress in sputtered tantalum thin-film. *Appl. Surf. Sci.* **2016**, *371*, 571–575. [CrossRef]
51. Zhou, Y.M.; Xie, Z.; Xiao, H.N.; Hu, P.F.; He, J. Effects of deposition parameters on tantalum films deposited by direct current magnetron sputtering. *J. Vac. Sci. Technol. A: Vacuum, Surfaces, Films* **2009**, *27*, 109–113. [CrossRef]
52. Bernoulli, D.; Müller, U.; Schwarzenberger, M.; Hauert, R.; Spolenak, R. Magnetron sputter deposited tantalum and tantalum nitride thin films: An analysis of phase, hardness and composition. *Thin Solid Films* **2013**, *548*, 157–161. [CrossRef]
53. Zhou, Y.; Xie, Z.; Xiao, H.; Hu, P.; He, J. Effects of deposition parameters on tantalum films deposited by direct current magnetron sputtering in Ar–O2 mixture. *Appl. Surf. Sci.* **2011**, *258*, 1699–1703. [CrossRef]
54. Zhou, Y.; Xie, Z.; Ma, Y.; Xia, F.; Feng, S. Growth and characterization of Ta/Ti bi-layer films on glass and Si (111) substrates by direct current magnetron sputtering. *Appl. Surf. Sci.* **2012**, *258*, 7314–7321. [CrossRef]
55. Navid, A.; Chason, E.; Hodge, A. Evaluation of stress during and after sputter deposition of Cu and Ta films. *Surf. Coatings Technol.* **2010**, *205*, 2355–2361. [CrossRef]
56. Myers, S.; Lin, J.; Souza, R.; Sproul, W.D.; Moore, J.J. The β to α phase transition of tantalum coatings deposited by modulated pulsed power magnetron sputtering. *Surf. Coatings Technol.* **2012**, *214*, 38–45. [CrossRef]
57. Cacucci, A.; Loffredo, S.; Potin, V.; Imhoff, L.; Martin, N. Interdependence of structural and electrical properties in tantalum/tantalum oxide multilayers. *Surf. Coatings Technol.* **2012**, *227*, 38–41. [CrossRef]
58. Navid, A.; Hodge, A. Nanostructured alpha and beta tantalum formation—Relationship between plasma parameters and microstructure. *Mater. Sci. Eng. A* **2012**, *536*, 49–56. [CrossRef]
59. Navid, A.; Hodge, A. Controllable residual stresses in sputtered nanostructured alpha-tantalum. *Scr. Mater.* **2010**, *63*, 867–870. [CrossRef]
60. Zabolotnyi, V.T. *Ion Intermixing in Solids*; MGIEM (TU): Moscow, Russian, 1997; p. 62 p.
61. Kuz'michev, A.I. *Magnetron Spattering Systems. Book 1. Introduction Into Physics And Technique Of Magnetron Scattering*; Avers: Kiev, Ukraine, 2008; p. 244 p.
62. Poate, J.M.; Foti, G.; Jacobson, D.C. *Surface Modification and Alloying by Laser*; Ion and Electron Beams: Plenum, NY, USA, 1983; p. 424 p.
63. Dorranian, D.; Solati, E.; Hantezadeh, M.; Ghoranneviss, M.; Sari, A. Effects of low temperature on the characteristics of tantalum thin films. *Vacuum* **2011**, *86*, 51–55. [CrossRef]
64. Maeng, S.; Axe, L.; Tyson, T.; Gladczuk, L.; Sosnowski, M. Corrosion behaviour of magnetron sputtered α- and β-Ta coatings on AISI 4340 steel as a function of coating thickness. *Corros. Sci.* **2006**, *48*, 2154–2171. [CrossRef]
65. Su, Y.; Huang, W.; Zhang, T.; Shi, C.; Hu, R.; Wang, Z.; Cai, L. Tribological properties and microstructure of monolayer and multilayer Ta coatings prepared by magnetron sputtering. *Vacuum* **2021**, *189*, 110250. [CrossRef]
66. Niu, Y.; Chen, M.; Wang, J.; Yang, L.; Guo, C.; Zhu, S.; Wang, F. Preparation and thermal shock performance of thick α-Ta coatings by direct current magnetron sputtering (DCMS). *Surf. Coatings Technol.* **2017**, *321*, 19–25. [CrossRef]
67. Zhang, M.; Yang, B.; Chu, J.; Nieh, T. Hardness enhancement in nanocrystalline tantalum thin films. *Scr. Mater.* **2006**, *54*, 1227–1230. [CrossRef]
68. Zhang, Y.; Wei, Q.; Niu, H.; Li, Y.; Chen, C.; Yu, Z.; Bai, X.; Zhang, P. Formation of nanocrystalline structure in tantalum by sliding friction treatment. *Int. J. Refract. Met. Hard Mater.* **2014**, *45*, 71–75. [CrossRef]
69. Shankar, V.; Mariappan, K.; Nagesha, A.; Reddy, G.P.; Sandhya, R.; Mathew, M.; Jayakumar, T. Effect of tungsten and tantalum on the low cycle fatigue behavior of reduced activation ferritic/martensitic steels. *Fusion Eng. Des.* **2012**, *87*, 318–324. [CrossRef]
70. Zhang, Y.; Zhang, X.; Wang, G.; Bai, X.; Tan, P.; Li, Z.; Yu, Z. High strength bulk tantalum with novel gradient structure within a particle fabricated by spark plasma sintering. *Mater. Sci. Eng. A* **2011**, *528*, 8332–8336. [CrossRef]
71. Silva, R.A.; Silva, I.P.; Rondot, B. Effect of Surface Treatments on Anodic Oxide Film Growth and Electrochemical Properties of Tantalum used for Biomedical Applications. *J. Biomater. Appl.* **2006**, *21*, 93–103. [CrossRef]
72. Chakraborty, B.; Halder, S.; Maurya, K.; Srivastava, A.; Toutam, V.; Dalai, M.; Sehgal, G.; Singh, S. Evaluation of depth distribution and characterization of nanoscale Ta/Si multilayer thin film structures. *Thin Solid Films* **2012**, *520*, 6409–6414. [CrossRef]
73. Colin, J.J.; Abadias, G.; Michel, A.; Jaouen, C. On the origin of the metastable β-Ta phase stabilization in tantalum sputtered thin films. *Acta Mater.* **2017**, *126*, 481–493. [CrossRef]
74. Nasakina, E.O.; Sevost'Yanov, M.A.; Mikhailova, A.B.; Gol'Dberg, M.A.; Demin, K.Y.; Kolmakov, A.G.; Zabolotnyi, V.T. Preparation of a nanostructured shape-memory composite material for biomedical applications. *Inorg. Mater.* **2015**, *51*, 400–404. [CrossRef]
75. Nasakina, E.; A Sevostyanov, M.; Mikhaylova, A.B.; Baikin, A.S.; Sergienko, K.V.; Leonov, A.V.; Kolmakov, A.G. Formation of alpha and beta tantalum at the variation of magnetron sputtering conditions. *IOP Conf. Series: Mater. Sci. Eng.* **2016**, *110*, 012042. [CrossRef]

Disclaimer/Publisher's Note: The statements, opinions and data contained in all publications are solely those of the individual author(s) and contributor(s) and not of MDPI and/or the editor(s). MDPI and/or the editor(s) disclaim responsibility for any injury to people or property resulting from any ideas, methods, instructions or products referred to in the content.

Article

Enhancement of Power Conversion Efficiency with Zinc Oxide as Photoanode and *Cyanococcus*, *Punica granatum* L., and *Vitis vinifera* as Natural Fruit Dyes for Dye-Sensitized Solar Cells

Ili Salwani Mohamad [1,2], Mohd Natashah Norizan [1,2,*], Norsuria Mahmed [2,3], Nurnaeimah Jamalullail [1], Dewi Suriyani Che Halin [2,3], Mohd Arif Anuar Mohd Salleh [2,3], Andrei Victor Sandu [4,*], Madalina Simona Baltatu [4,*] and Petrica Vizureanu [4,5]

1. Faculty of Electronic Engineering & Technology, Universiti Malaysia Perlis (UniMAP), Arau 02600, Malaysia
2. Centre of Excellence Geopolymer and Green Technology (CEGeoGTech), Universiti Malaysia Perlis (UniMAP), Arau 02600, Malaysia
3. Faculty of Chemical Engineering & Technology, Universiti Malaysia Perlis (UniMAP), Arau 02600, Malaysia
4. Department of Technologies and Equipments for Materials Processing, Faculty of Materials Science and Engineering, Gheorghe Asachi Technical University of Iaşi, Blvd. Mangeron, No. 51, 700050 Iasi, Romania
5. Technical Sciences Academy of Romania, Dacia Blvd 26, 030167 Bucharest, Romania
* Correspondence: mohdnatashah@unimap.edu.my (M.N.N.); sav@tuiasi.ro (A.V.S.); madalina-simona.baltatu@academic.tuiasi.ro (M.S.B.)

Abstract: Ruthenium N719 is a well-known material used as the dye in commercial dye-sensitized solar cell (DSSC) devices. However, it poses risks to human health and the environment over time. On the other hand, titanium dioxide (TiO_2) has low electron mobility and high recombination losses when used as a photoanode in this photovoltaic technology device. In addition, using Ruthenium as the dye material harms the environment and human health. As an alternative sensitizer to compensate Ruthenium on two different photoanodes (TiO_2 and ZnO), we constructed DSSC devices in this study using three different natural dyes (blueberry, pomegranate, and black grape). In good agreement with the anthocyanin content in the fruits, black grape, with the highest anthocyanin content (450.3 mg/L) compared to other fruit dyes (blueberry—386.6 mg/L and pomegranate—450.3 mg/L), resulted in the highest energy conversion efficiency (3.63%) for the natural dye-based DSSC. Furthermore, this research proved that the electrical performance of natural dye sensitizer in DSSC applications with a ZnO photoanode is better than using hazardous Ru N719 dye with a TiO_2 photoanode owing to the advantage of high electron mobility in ZnO.

Keywords: dye-sensitized solar cells; natural dye; zinc oxide; titanium oxide; anthocyanin; renewable energy; solar cells

1. Introduction

The third generation of solar cell technology, known as the dye-sensitized solar cell (DSSC) or Gratzel Cell, uses the photovoltaic effect to turn solar energy (photons) into electricity [1]. Due to its low cost of production, straightforward fabrication method, and efficiency performance comparable to other photovoltaic technology devices, this cell has attracted much research since Gratzel and O'Regan first developed it in 1991 [2]. Resembling the photosynthesis process, in DSSC, a dye that serves as the absorber layer (such as chlorophyll) absorbs photons and excites the electrons in the valence band to the conduction band. The electrons are then transferred from the conduction band to the outer circuit via metal oxide material, and a redox process completes the circuit at the counter electrode [3].

DSSC is a sandwich cell, as shown in Figure 1, that consists of ITO-coated glass as the electrodes and mechanical support [4], a wide bandgap mesoporous oxide layer photoanode [5] that serves to anchor the molecules of the dye sensitizer and acts as an electron transport medium, a dye which responsible for absorption of solar radiation for electron generation [6], and an electrolyte based on iodide/triiodide redox system between photosensitized photoanode and transparent conducting counter electrode [7]. Technically, the DSSC's working principle can be simplified into the following [5];

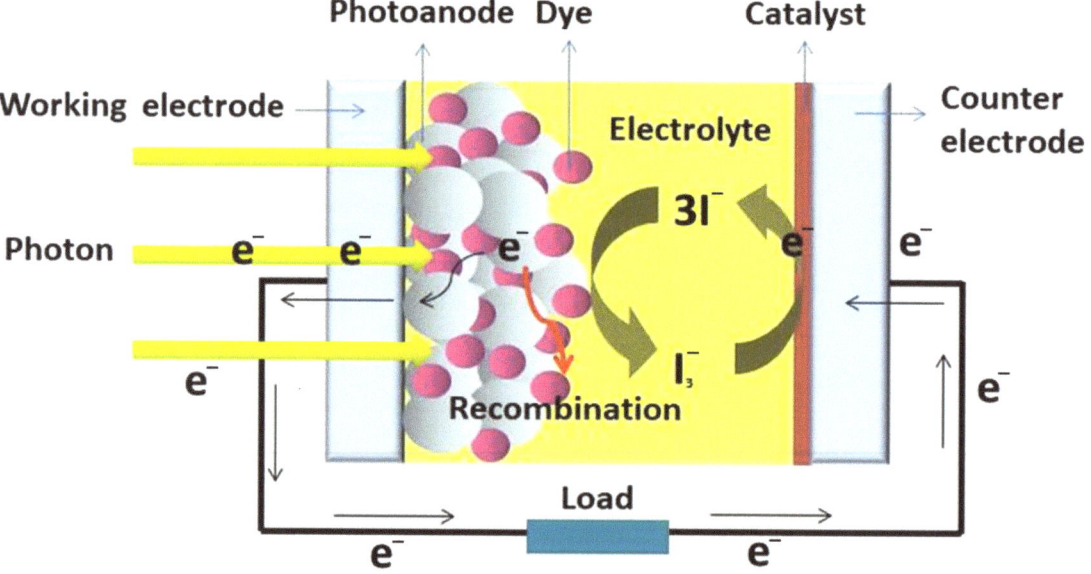

Figure 1. An electrical and chemical reaction in a DSSC structure during photon absorption.

A dye molecule (S) is struck by photon energy ($h\nu$), which causes it to enter the excited state (S^*).

$$S + h\nu \rightarrow S^* \tag{1}$$

When an electron escapes from a dye molecule and enters a photoanode nanostructure, the molecule develops a hole (S^+).

$$S^* \rightarrow S^+ + e^- \tag{2}$$

The redox system regenerates the dye molecule.

$$S^+ + 3/2 I^- \rightarrow S + 1/2 I_3^- \tag{3}$$

Electrons passing through the load regenerate the redox couple at the counter electrode.

$$I_3^- + 2e^- \rightarrow 3I^- \tag{4}$$

The process will repeat in the presence of sunlight.

The commercialized DSSC utilizes TiO_2 and metal complex N719, *Ruthenium* (Ru), as the photoanode and dye sensitizer. Due to their advantageous photoelectrochemical characteristics and great stability in the oxidized state, ruthenium complexes have attracted particular interest as photosensitizers in DSSC applications [8]. Modern DSSCs with ruthenium (II)- polypyridyl complexes (N719) as the active materials currently have total power conversion efficiencies of about 11% under AM1.5G light conditions [9]. Besides the extensive range of absorption from the visible to the near-infrared (NIR) spectrum, careful study of the HOMO and LUMO energy levels can be used to adjust the ruthenium polypyridyl complexes' absorption spectra, which may increase the performance of the cell [10].

Unfortunately, besides the advantages listed above, the DSSC structure developed using TiO$_2$ and Ru N719 have drawbacks that impact the device's performance and are carcinogenic. TiO$_2$ has low electron mobility (0.1–4 cm^{2}V^{-1}s^{-1}) [11], which causes the DSSC to have a high recombination rate [12]. This limitation of TiO$_2$ reduces the number of carriers transported to the front contact and disturbs the overall cell efficiency performance. The red line in Figure 1 shows the electron recombination process phenomena, which results in a loss in energy conversion efficiency; it occurs when the electrons in the TiO$_2$ structure recombine with the oxidized dye molecules and electrolyte rather than being transported to the anode and cathode terminals. In addition, because the TiO$_2$ phase is metastable and can change into either brookite, anatase, or rutile under specific conditions such as temperature, the nanostructure of TiO$_2$ is also restricted to being synthesized [13]. Ru, on the other hand, is a high-cost material [14], rare [15], and complex to be synthesized [16]. Worryingly, Ru sensitizer material has been reported as toxic and carcinogenic, harming human health [17]. It has been reported that the basic requirements for a good sensitizer are (a) a strong dye attached to a semiconductor material, (b) a broad absorption spectrum, and (c) being able to inject the electron into semiconductor materials [18].

In this work, we explored the possibility of using ZnO, which has higher electron mobility compared to TiO$_2$, as the photoanode for the test cells using three different organic dyes: *Cyanococcus* (Blueberry), *Punica granatum* L. (Pomegranate), and *Vitis vinifera* (Black Grape). Organic dyes were synthesized using chemical mixing, and the doctor blade technique was used to layer the photoanode material on the front contact of the respective devices. The properties of the photoanodes (morphology and phase) and the electrical performance of the DSSC devices with various photoanode materials and dyes used were reviewed, measured, and studied. Our previous study on natural pigments shows that photosensitizer dyes [19] such as chlorophyll, anthocyanin, and betalain extracted from various plants' leaves, flowers, and fruits have proved their efficacy as a good dye selection for organic-based DSSCs.

2. Materials and Methods

2.1. Preparation of TiO$_2$ and ZnO Photoanode

Two transparent conductive oxide (TCO) coated glasses were cleaned as in [20]. Following this, 1 g of 99% anatase TiO$_2$ powder and 1.5 mL of acetic acid solution (0.1 M) were combined in a pestle and mortar to create the TiO$_2$ solution. A solution of 0.1 M acetic acid was prepared beforehand by mixing 10 mL of deionized water with 58 µL of 99.99% acetic acid. A white soupy solution was then obtained by stirring the acetic acid solution and photoanode powder together. In the other pestle and mortar, 1 g of ZnO powder (98%) with 6 mL of ethanolic solution (0.01 M) and 0.5 mL of nitric acid solution (0.1 M) were mixed, creating a ZnO photoanode soupy solution. Ethanolic 0.01 M solutions were prepared beforehand by mixing absolute ethanol with DI water in a ratio of 7:3 to prevent photoanode paste from drying before applying it to the ITO-coated glass. Meanwhile, the 0.1 M nitric acid solution comprised 137 µL of 65% nitric acid base and 20 mL of DI water.

TiO$_2$ and ZnO photoanodes were prepared using the doctor blade technique with a 1.0×1.5 cm^2 active area on the conductive side of the ITO-coated glass, as shown in Figure 2. The deposited photoanode material was then annealed for 1 h at 450 °C in the ambient condition furnace.

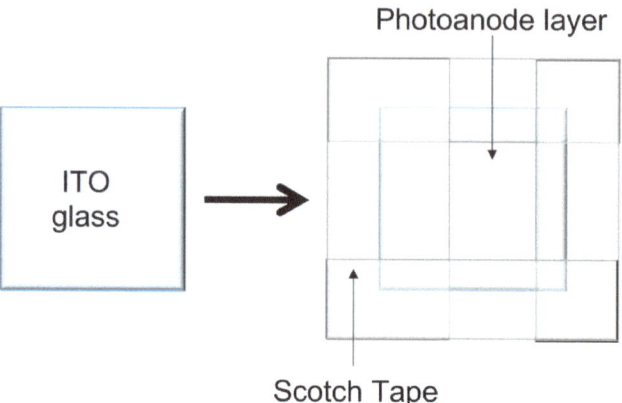

Figure 2. Photoanode semiconductor material deposition using the doctor blade method.

2.2. Preparation of Dye Sensitizers & Dye Coating Process

Due to the different environments and facilities of the laboratory where the DSSCs were fabricated, the efficiency of conventional DSSCs obtained in this study may differ from the literature. In order to determine the effect of photoanode and dye variations on the theoretical support hypothesis, a baseline cell of TiO_2 with *Ruthenium* (Ru N719) dyes was constructed prior to the implementation of any improvements for comparison and justification. Ru N719 dye was prepared as [20] for the baseline sample. Natural dye extracts from blueberry, black grape, and pomegranate were then prepared by crushing 50 g of selected fruits using a mortar. The extracts were filtered and diluted in 20 mL of ethanol and the solutions in Figure 3 were obtained. According to previous research [21], the presence of ethanol in the dye solution can stabilize the acidity of the fruit dyes to prevent the photoanode from dissolving, which could reduce the DSSC's ability to convert sunlight into energy.

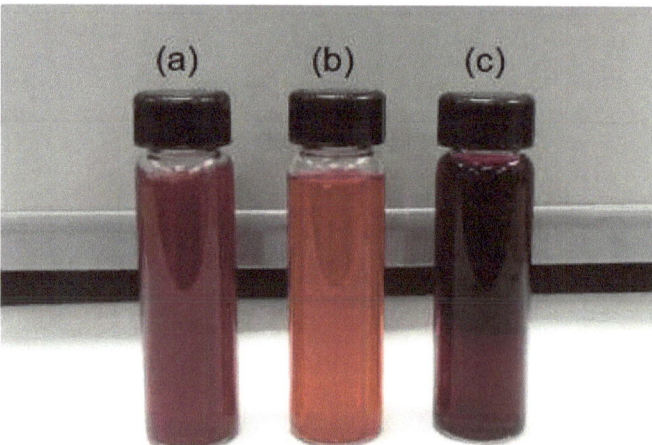

Figure 3. (**a**) black grape; (**b**) pomegranate, and (**c**) blueberry dye solutions.

As shown in Figure 4, the prepared photoanodes were immersed in the dye solution for two hours in a sealed petri dish.

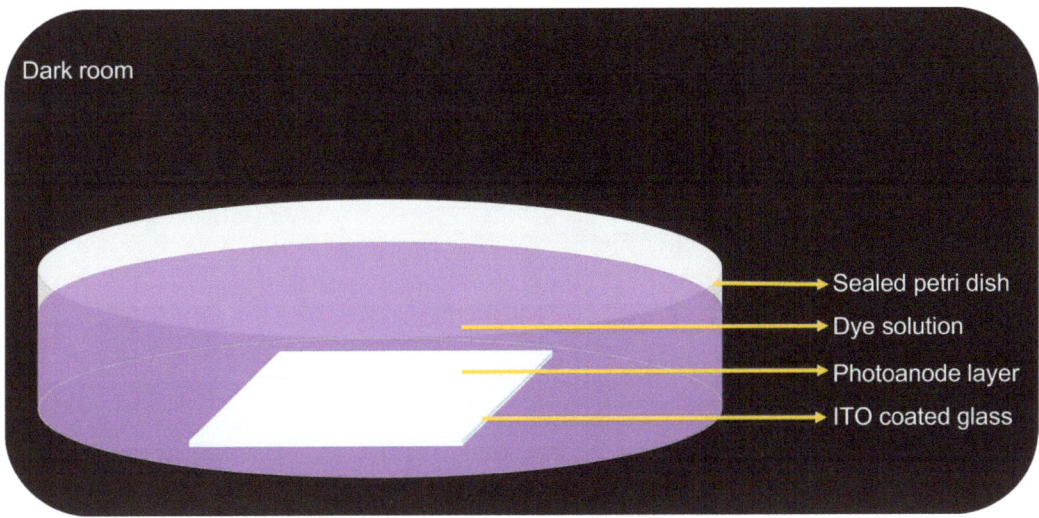

Figure 4. Photoanode dye coating process in dark ambient.

The longer time taken for this step may cause peel-off at the photoanode layer due to the acidity factor in the dyes. The samples were placed in a dark, enclosed area in a sealed petri dish covered in aluminum foil. Figure 5 shows the working electrode (photoanode) before and after the dye soaking process.

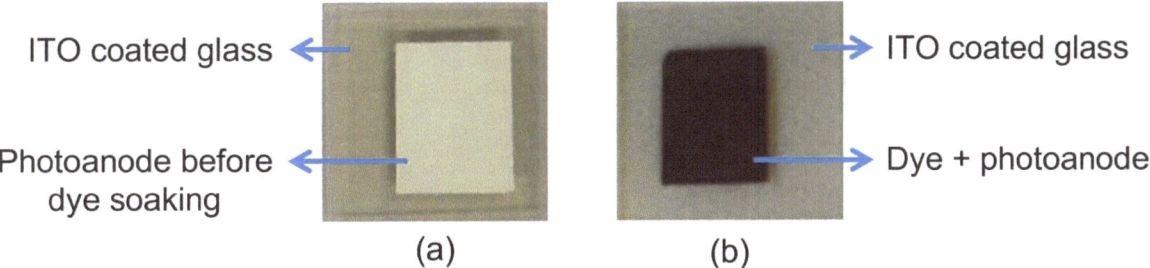

Figure 5. The working electrode (**a**) before the dye soaking process and (**b**) after 2 h of the dye coating process.

2.3. Preparation of Counter Electrode

On the ITO-coated glass, two holes with diameters of 1 mm and 0.5 mm were drilled. The 1-mm-size hole was created as the route for the electrolyte injection because the final DSSC will be sealed at the edge to prevent air from reducing the efficiency of the cell. The 0.5 mm hole served as a passageway for the trapped air in the DSSC structure to exit so that the electrolyte solution could fill the space between the two electrodes. Using the candle soot, the carbon layer with 1.0×1.5 cm^2 active area was deposited on the ITO-coated glass's conductive side Finally, the excess carbon layer was rubbed using a cotton bud. Figure 6 shows the counter electrode's final product.

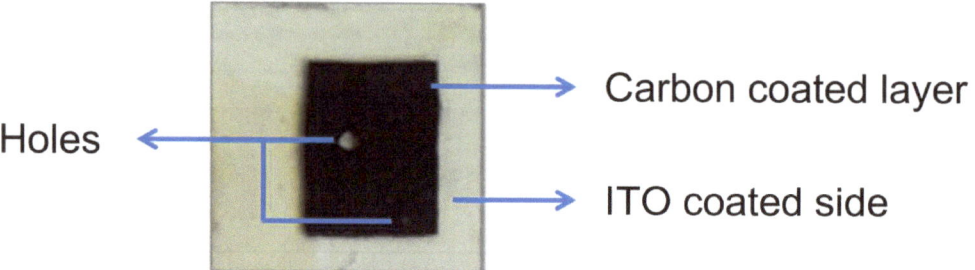

Figure 6. Carbon-coated counter electrode.

2.4. Dye-Sensitized Solar Cells (DSSC) Assembly

DSSC was constructed by sandwiching a photoanode that was soaked in the dye with the counter electrode sealed with parafilm at the side (heated at 80 °C for 30 s with 500 g force applied on it to ensure that the DSSC was completely sealed). A potassium iodide (KI) solution that acts as an electrolyte was dropped into the sample through the hole at the counter electrode. The hole was then sealed by using scotch tape. The sample was then cooled at room temperature. This step is very important to prevent the electrolyte from leaking.

The electrical properties and energy conversion efficiency of the constructed DSSC were examined using a solar simulator (Keithley SMU 2450, Keithley, Tektronix, Beaverton, OR, USA). The DSSC was subjected to 1000 W/m² of light produced by a xenon lamp power supply (XPS-1600, Solar Light Inc., Glenside, PA, USA), simulating actual solar energy during the electrical testing. The current density–voltage (J–V) characteristics can be used to determine energy conversion efficiency. The efficiency of a solar cell is calculated from Equation (5) [22], where P_{in} is the input power from the sun at AM1.5G with a value of 1000 W/m². This mathematical relation shows that J_{sc}, V_{oc}, and FF are the three most important indicators that affect the output performance of a photovoltaic cell.

$$\eta = \frac{J_{sc} \times V_{oc} \times FF}{P_{in}} = \frac{J_{sc} \times V_{oc} \times FF}{I_{max} \times V_{max}} \quad (5)$$

where

η = efficiency
J_{sc} = short circuit current
V_{oc} = open circuit voltage
FF = fill factor
P_{in} = input power
I_{max} = maximum current
V_{max} = maximum voltage

The factors affecting the energy conversion efficiency of the DSSC were examined through the crystallite size and structure of the photoanode, which were seen using an X-ray diffraction instrument (D2 Phaser, Bruker Ltd., Billerica, MA, USA). A scanning electron microscope (SEM, JOEL JSM-6010LV, JOEL Ltd., Tokyo, Japan) was also used to examine the surface morphology of each synthesized photoanode. ImageJ was used to determine the average particle size. The ability of the natural fruit dyes from blueberry, pomegranate, and black grape to absorb light was investigated using UV-visible spectroscopy (Lambda 950, Perkin Elmer, Inc., Waltham, MA, USA).

3. Results and Discussions

Based on the electrical performance testing for the laboratory baseline cell, the TiO_2-based photoanode with inorganic Ru N719 dye recorded 1.31% compared to 2.22% for the ZnO-based photoanode DSSC with the same dye, as stated in Table 1. Compared to our previous research in [20], 2 h of soaking time for the photoanode in the Ru N719 dye in this work has improved the electrical performance of both the TiO_2 and ZnO photoanode. TiO_2-based photoanode DSSC portrayed better J_{sc} and a slightly higher fill factor than ZnO-based DSSC. However, in contrast, the ZnO photoanode shows a significant increment in V_{oc}, leading to better efficiency.

Table 1. Electrical performance for the TiO_2 and ZnO Baseline Cell with Ru N719 dye.

Dye	Photoanode	J_{sc} (mA/cm^2)	V_{oc} (V)	Fill Factor (FF)	η (%)
Ru N719	TiO_2	7.47	0.36	0.59	1.31
	ZnO	5.18	0.79	0.54	2.22

The average crystallite size for TiO_2 and ZnO calculated using the Scherrer equation based on XRD evaluation were 18.76 nm and 19.30 nm, respectively. These findings validated that the ZnO photoanode measured energy conversion efficiency, which was 69.5% better than the TiO_2 photoanode containing Ru N719 dye.

The ZnO photoanode's magnificent improvement can be attributed to the morphological features in the photoanode layer, according to the results of the XRD analysis and SEM surface morphology identification in our earlier work. The ZnO photoanode layer's rectangular nanostructures, as seen in Figure 7, have a lot of surface area for the dye to adhere to, enabling the DSSC to absorb more energy [23].

Figure 7. Surface morphology of SEM for (**a**) TiO2 and (**b**) ZnO [20].

Furthermore, the larger ZnO photoanode structure, which consists of rectangular-shaped nanostructures, eventually creates a light-scattering layer and aids in enhancing the DSSC's capacity to harvest energy by reflecting light that passes through the photoanode layer so that the dye molecules can reabsorb it, as demonstrated by the analogy in Figure 8. This finding is well-concordant with earlier research [24].

The higher energy conversion efficiency of the ZnO photoanode is also supported by the material's chemical characteristics, which include high electron mobility (200 cm^{2}V^{-1}s^{-1}) and large free excitation binding energy (60 meV) [25]. Thus, it was proved that ZnO has a higher energy conversion efficiency for a semiconductor photoanode than a TiO_2 photoanode.

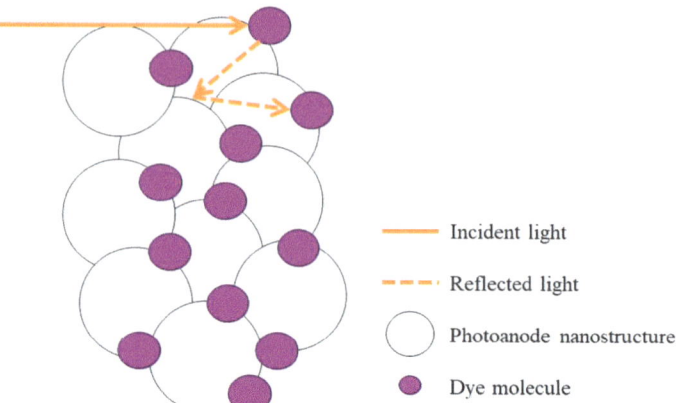

Figure 8. Light scattering effect on large particle size semiconductor photoanode material.

The electrical performance of DSSCs using various natural fruit dyes containing anthocyanin pigments is summarized in Table 2. The highest energy conversion efficiency for natural dye-based DSSC is found in the ZnO photoanode-based device with black grape dye (3.63%). We can observe that for all organic dyes, most ZnO-based DSSC was reported to give higher V_{oc} than TiO$_2$-based DSSC. These findings are consistent with the hypothesis that materials with higher bandgap values have higher V_{oc} values, as in Equation (6) [26], which may be associated with slower recombination and better electron collection. This argument is backed by the characteristics of ZnO material, which is known to have better electron mobility than TiO$_2$ material, which further enhances electron transport and makes charge carrier separation easier. Different organic dyes used do not seem to affect V_{oc} significantly, but different photoanodes do.

$$qV_{OC} = \left(1 - \frac{T}{T_{sun}}\right) - kT\left[\ln\left(\frac{\Omega_{emit}}{\Omega_{sun}}\right) + \ln\left(\frac{4n^2}{I}\right) - \ln(QE)\right] \quad (6)$$

Table 2. Electrical performance of the constructed DSSC using various photoanodes and dye types.

Dye	Photoanode	J_{sc} (mA/cm^2)	V_{oc} (V)	FF	η (%)
Blueberry	TiO$_2$	0.82	1.18	0.40	0.39
	ZnO	7.12	0.79	0.50	2.81
Pomegranate	TiO$_2$	5.86	0.35	0.45	0.93
	ZnO	7.12	0.71	0.60	3.03
Black grape	TiO$_2$	6.69	0.52	0.30	1.05
	ZnO	9.72	0.73	0.51	3.63

In order to study the separation efficiency of photogenerated electrons and holes, a room temperature photoluminescence (PL) spectroscopy evaluation of the photoanodes (and dyes) is needed, which was not covered in this work. However, it is reported that the decrease in PL intensity indicates efficient electron-hole separation and long-lived carriers, which may effectively reduce the recombination of electrons and holes.

The highlight of this study is the utilization of ZnO-based photoanode soaked with different organic dye materials to see the effect and impact of the interaction between the anthocyanin level of the dyes (compared to the commercial Ru N719 dye) with ZnO semiconductor material photoanode to the performance of DSSC. The unique, organic dye materials selected were specifically from blueberry, pomegranate, and black grape, which—reflecting different anthocyanin levels (with different carboxylic group content)—produced a direct proportional relationship between the anthocyanin content level and the electrical performance of the DSSC. With regards to our previous review on the natural pigment absorber material for photon absorption and efficient electrical conversion in [19], it can be seen that the results obtained from this study are better, as J_{sc} = 9.72 mA/cm^{-2}, V_{oc} = 0.73 V, FF = 0.51, and efficiency = 3.63%. This result is not solely contributed by the absorber material used; the photoanode semiconductor material used as the electron transporter also significantly contributes to this result.

Figure 9 shows the trends in the short circuit current density-open circuit voltage (J_{sc}-V_{oc}) characteristics for each tested sample. Overall, it can be concluded that all ZnO-based photoanode cells are performing better than TiO$_2$-based cells regardless of organic dyes tested with more convincing J–V curve shape patterns. These results support ZnO's role as a photoanode with higher electron mobility properties as a better solution for electron transport to the front contact of the device than TiO$_2$. The shorter time for the electron to be transported to the electrode minimizes the recombination process, thus enhancing efficiency. Black grape recorded the highest efficiency with the best J_{sc} among all cells tested compared to the other natural dyes used. As the dye is important to function as the photon absorber and electron generator for the DSSC cell, black grape proves that a high anthocyanin content in a material is very important for the role. The advantage of using ZnO as a photoanode compared to TiO$_2$ in terms of output performance also has been proved in our previous research [20] utilizing Ru N719 dye as the sensitizer.

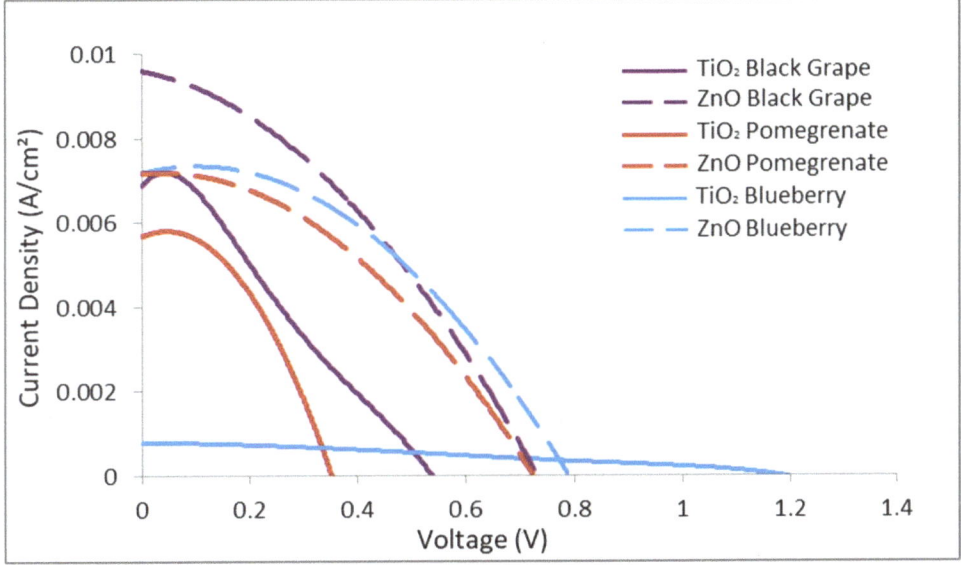

Figure 9. J–V curve for DSSCs with different types of photoanodes and dye sensitizers.

According to the trend in Figure 9, even though all DSSCs tested show a valid J–V curve for a practical photovoltaic (PV) device compared to the best DSSC cell reported by NREL, it has a lower fill factor, which can be seen at the area under the graph. The efficiency loss is partly due to the increased series resistance [27]. Series resistance can be caused by the unideal series connection of the electrodes, interfacial resistance between electrode and TCO, and electrolyte [28]. Series resistance also can be obtained from the following formula [29];

$$R_s = R_{TCO} + R_{CT} + R_{diff}(I_3^-) \qquad (7)$$

where;
R_{TCO} = substrate resistance
$R_{diff}(I_3^-)$ = diffusion impedance of I_3^- ions in the electrolyte
R_{CT} = charge transfer resistance

Further work is needed to understand the detailed causes of loss due to series resistance to improve future device stability. A good solar cell must have near-zero series resistance and extremely high shunt resistance [30]. The absorption spectra of the dye extracts made from blueberry, black grape, and pomegranate are represented in Figure 10. The maximum absorption peaks (max) for blueberry, pomegranate, and black grape were seen at 568 nm, 539 nm, and 537 nm, respectively, corresponding to the anthocyanin pigments, which also matched the other study [31]. All anthocyanin dyes absorbed photons in UV (~350 nm) and visible light wavelengths.

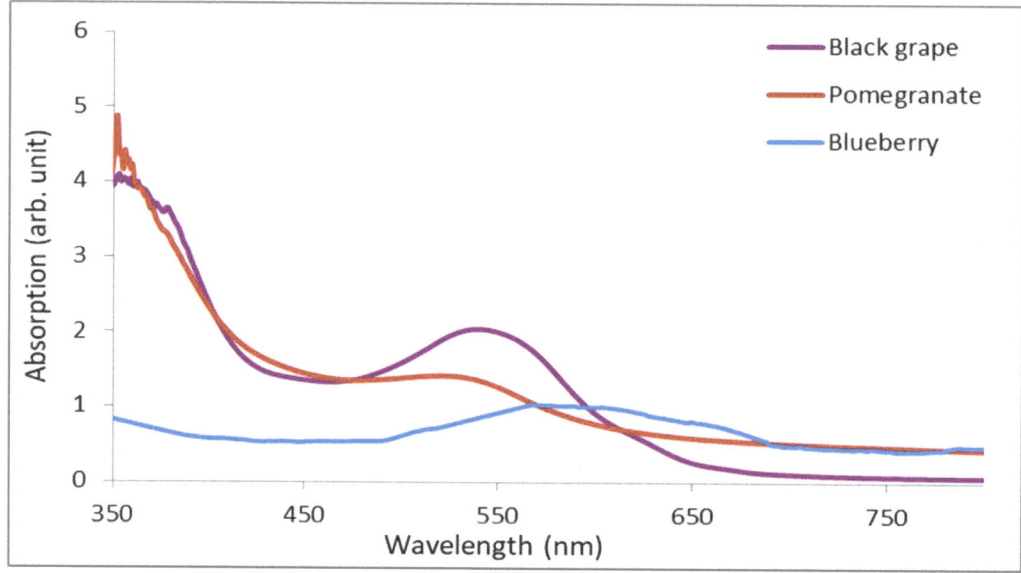

Figure 10. The absorption spectrum of anthocyanin fruit dyes.

The observed absorption peaks are congruent with the anthocyanin content. Black grape dye, having the highest anthocyanin content (450.3 mg/L) [32], shows high absorption and an intense peak compared to pomegranate (409.4 mg/L) [33] and blueberry dye (386.6 mg/L) [34]. Clearly, the trend of the TiO_2 blueberry curve differs from that of the other curves in Figures 9 and 10. This is because the blueberry dye contains less anthocyanin than other organic dyes, which produces fewer electrons during the photon-to-electrical conversion when the cell is illuminated by sunlight. Therefore, based on the pattern curve, this cell has very high series resistance and very low shunt resistance. This research showed how fruit dyes contribute to a better DSSC's energy conversion efficiency due to anthocyanin contents in the dye. The anthocyanin dye pigments are attached to semiconductor photoanode structures thanks to the presence of carbonyl and hydroxyl (COOH) group chains that serve as anchorage agents. The number of dye pigments attached to the semiconductor structure will determine how much sunlight can be absorbed by the DSSC.

The energy bandgap of the three anthocyanin fruit dyes was calculated by using equation $E = hc/\lambda$; where h = Planck's constant ($h = 6.63 \times 10^{-34}$), c = speed of light ($c = 3 \times 10^8$ ms^{-1}), and λ_{max} = wavelength of the maximum peak of UV–visible absorption spectrum [35]. Table 3 summarizes the energy bandgap of the anthocyanin fruit dyes:

blueberry (2.18 eV), pomegranate (2.30 eV), and black grape (2.32 eV). It can be observed that λ_{max} is inversely proportional to the energy bandgap of the anthocyanin dye. This is because a lower E_g material encounters a thermalization issue where the excess energy absorbed may turn into heat, leading to efficiency loss [36]. This justification supports our observation that the electrical performance of black grape dye with E_g 2.32 eV gives an energy conversion efficiency of 3.63% higher than blueberry and pomegranate dye.

Table 3. Energy bandgap values of blueberry, pomegranate, and black grape dye.

Dye	λ_{max} (nm)	The Energy Band Gap (eV)
Blueberry	568	2.18
Pomegranate	539	2.30
Black grape	537	2.32

4. Conclusions

In this work, the performance of DSSC with organic and inorganic dyes has been examined in relation to the use of ZnO and TiO_2 as photoanodes. The ZnO photoanode has shown excellent performance due to having a bigger average crystallite and particle size in a rectangular shape compared to TiO_2 material, which helps to provide the light scattering layer and improves the light-harvesting ability of the DSSC. In addition, ZnO, with the advantage of high electron mobility (200 $cm^2V^{-1}s^{-1}$) and large free excitation binding energy (60 meV), reduces the recombination loss in the cell. Based on three organic dyes selected for this research, black grapefruit (anthocyanin content 450.3 mg/L) obtained the highest efficiency compared to blueberry and pomegranate. Furthermore, this work proved that power conversion efficiency is directly proportional to the anthocyanin content level in dye material for organic dyes.

5. Patents

The following code, PI2021004374, which corresponds to Malaysia, the country of origin of the invention, has been registered for patents resulting from the work described in this manuscript.

Author Contributions: Conceptualization, M.N.N. and I.S.M.; methodology, N.J., A.V.S. and I.S.M.; software, M.N.N.; validation, N.M., P.V. and D.S.C.H.; formal analysis, I.S.M., M.S.B. and N.M.; investigation, N.J., P.V., I.S.M. and M.N.N.; resources, N.M.; data curation, I.S.M. and A.V.S.; writing—original draft preparation, N.J.; writing—review and editing, I.S.M., M.S.B., M.N.N. and D.S.C.H.; visualization, M.N.N. and M.S.B.; supervision, I.S.M., M.N.N., A.V.S and N.M.; project administration, I.S.M., M.N.N., P.V. and M.A.A.M.S.; funding acquisition, M.N.N. and M.A.A.M.S. All authors have read and agreed to the published version of the manuscript.

Funding: This research was funded by the Ministry of Higher Education of Malaysia (MoHE) and Universiti Malaysia Perlis (UniMAP) under the Fundamental Research Grant Scheme (FRGS) (Grant number: FRGS/1/2020/TK0/UNIMAP/02/35) and by Gheorghe Asachi Technical University of Iasi—TUIASI Romania, Scientific Research Funds, FCSU-2022.

Institutional Review Board Statement: Not applicable.

Informed Consent Statement: Not applicable.

Data Availability Statement: Not applicable.

Acknowledgments: Special thanks to the Centre of Excellence for Renewable Energy (CERE), and UniMAP for providing the characterization equipment.

Conflicts of Interest: The authors declare no conflict of interest.

References

1. O'Regan, B.; Grätzel, M. A Low-Cost, High-Efficiency Solar Cell Based on Dye-Sensitized Colloidal TiO2 Films. *Nature* **1991**, *353*, 737–740. [CrossRef]

2. Pham, H.D.; Yang, T.C.; Jain, S.M.; Wilson, G.J.; Sonar, P. Development of Dopant-Free Organic Hole Transporting Materials for Perovskite Solar Cells. *Adv. Energy Mater.* **2020**, *10*, 1903326. [CrossRef]
3. Zaine, S.N.A.; Mohamed, N.M.; Khatani, M.; Samsudin, A.E.; Shahid, M.U. Trap State and Charge Recombination in Nanocrystalline Passivized Conductive and Photoelectrode Interface of Dye-Sensitized Solar Cell. *Coatings* **2020**, *10*, 284. [CrossRef]
4. Hossain, M.K.; Pervez, M.F.; Mia, M.N.H.; Mortuza, A.A.; Rahaman, M.S.; Karim, M.R.; Islam, J.M.M.; Ahmed, F.; Khan, M.A. Effect of Dye Extracting Solvents and Sensitization Time on Photovoltaic Performance of Natural Dye Sensitized Solar Cells. *Results Phys.* **2017**, *7*, 1516–1523. [CrossRef]
5. Devadiga, D.; Selvakumar, M.; Shetty, P.; Santosh, M.S. Recent Progress in Dye Sensitized Solar Cell Materials and Photo-Supercapacitors: A Review. *J. Power Sources* **2021**, *493*, 229698. [CrossRef]
6. Devadiga, D.; Selvakumar, M.; Shetty, P.; Santosh, M.S. Dye-Sensitized Solar Cell for Indoor Applications: A Mini-Review. *J. Electron. Mater.* **2021**, *50*, 3187–3206. [CrossRef]
7. Zeng, K.; Tong, Z.; Ma, L.; Zhu, W.-H.; Wu, W.; Xie, Y. Molecular Engineering Strategies for Fabricating Efficient Porphyrin-Based Dye-Sensitized Solar Cells. *Energy Environ. Sci.* **2020**, *13*, 1617–1657. [CrossRef]
8. Kohle, O.; Grätzel, M.; Meyer, A.F.; Meyer, T.B. The Photovoltaic Stability of, Bis(Isothiocyanato)Rluthenium(II)-Bis-2, 2′bipyridine-4, 4′-Dicarboxylic Acid and Related Sensitizers. *Adv. Mater.* **1997**, *9*, 904–906. [CrossRef]
9. Qin, Y.; Peng, Q. Ruthenium Sensitizers and Their Applications in Dye-Sensitized Solar Cells. *Int. J. Photoenergy* **2012**, *2012*, 291579. [CrossRef]
10. Tomar, N.; Agrawal, A.; Dhaka, V.S.; Surolia, P.K. Ruthenium Complexes Based Dye Sensitized Solar Cells: Fundamentals and Research Trends. *Sol. Energy* **2020**, *207*, 59–76. [CrossRef]
11. Chandiran, A.K.; Abdi-Jalebi, M.; Nazeeruddin, M.K.; Grätzel, M. Analysis of Electron Transfer Properties of ZnO and TiO_2 Photoanodes for Dye-Sensitized Solar Cells. *ACS Nano* **2014**, *8*, 2261–2268. [CrossRef] [PubMed]
12. Zhang, J.; Zhou, P.; Liu, J.; Yu, J. New Understanding of the Difference of Photocatalytic Activity among Anatase, Rutile and Brookite TiO_2. *Phys. Chem. Chem. Phys.* **2014**, *16*, 20382–20386. [CrossRef] [PubMed]
13. Byrne, C.; Fagan, R.; Hinder, S.; McCormack, D.E.; Pillai, S.C. New Approach of Modifying the Anatase to Rutile Transition Temperature in TiO_2 Photocatalysts. *RSC Adv.* **2016**, *6*, 95232–95238. [CrossRef]
14. Alhamed, M.; Issa, A.; Doubal, A. Studying of Natural Dyes Properties as Photo-Sensitizer for Dye Sensitized Solar Cells (DSSC). *J. Electron Devices* **2012**, *16*, 1370–1383.
15. Hamadanian, M.; Safaei-Ghomi, J.; Hosseinpour, M.; Masoomi, R.; Jabbari, V. Uses of New Natural Dye Photosensitizers in Fabrication of High Potential Dye-Sensitized Solar Cells (DSSCs). *Mater. Sci. Semicond. Process.* **2014**, *27*, 733–739. [CrossRef]
16. Kushwaha, R.; Srivastava, P.; Bahadur, L. Natural Pigments from Plants Used as Sensitizers for TiO 2 Based Dye-Sensitized Solar Cells. *J. Energy* **2013**, *2013*, 654953. [CrossRef]
17. Calogero, G.; Bartolotta, A.; Di Marco, G.; Di Carlo, A.; Bonaccorso, F. Vegetable-Based Dye-Sensitized Solar Cells. *Chem. Soc. Rev.* **2015**, *44*, 3244–3294. [CrossRef]
18. Yahya, M.; Bouziani, A.; Ocak, C.; Seferoğlu, Z.; Sillanpää, M. Organic/Metal-Organic Photosensitizers for Dye-Sensitized Solar Cells (DSSC): Recent Developments, New Trends, and Future Perceptions. *Dye. Pigment.* **2021**, *192*, 109227. [CrossRef]
19. Jamalullail, N.; Mohamad, I.S.; Norizan, M.N.; Baharum, N.A.; Mahmed, N. Short Review: Natural Pigments Photosensitizer for Dye-Sensitized Solar Cell (DSSC). In Proceedings of the 2017 IEEE 15th Student Conference on Research and Development (SCOReD), IEEE, Putrajaya, Malaysia, 13–14 December 2018; Volume 2018, pp. 344–349.
20. Jamalullail, N.; Smohamad, I.; Nnorizan, M.; Mahmed, N. Enhancement of Energy Conversion Efficiency for Dye Sensitized Solar Cell Using Zinc Oxide Photoanode. *IOP Conf. Ser. Mater. Sci. Eng.* **2018**, *374*, 012048. [CrossRef]
21. Ayalew, W.A.; Ayele, D.W. Dye-Sensitized Solar Cells Using Natural Dye as Light-Harvesting Materials Extracted from Acanthus Sennii Chiovenda Flower and Euphorbia Cotinifolia Leaf. *J. Sci. Adv. Mater. Devices* **2016**, *1*, 488–494. [CrossRef]
22. Sánchez-García, M.A.; Bokhimi, X.; Maldonado-Álvarez, A.; Jiménez-González, A.E. Effect of Anatase Synthesis on the Performance of Dye-Sensitized Solar Cells. *Nanoscale Res. Lett.* **2015**, *10*, 306. [CrossRef] [PubMed]
23. Al-Agel, F.A.; Shaheer Akhtar, M.; Alshammari, H.; Alshammari, A.; Khan, S.A. Solution Processed ZnO Rectangular Prism as an Effective Photoanode Material for Dye Sensitized Solar Cells. *Mater. Lett.* **2015**, *147*, 119–122. [CrossRef]
24. Son, M.-K.; Seo, H.; Kim, S.-K.; Hong, N.-Y.; Kim, B.-M.; Park, S.; Prabakar, K.; Kim, H.-J. Analysis on the Light-Scattering Effect in Dye-Sensitized Solar Cell According to the TiO 2 Structural Differences. *Int. J. Photoenergy* **2012**, *2012*, 480929. [CrossRef]
25. Mohamed, I.S.; Ismail, S.S.; Norizan, M.N.; Murad, S.A.Z.; Abdullah, M.M.A. ZnO Photoanode Effect on the Efficiency Performance of Organic Based Dye Sensitized Solar Cell. *IOP Conf. Ser. Mater. Sci. Eng.* **2017**, *209*, 012028. [CrossRef]
26. Polman, A.; Atwater, H.A. Photonic Design Principles for Ultrahigh-Efficiency Photovoltaics. *Nat. Mater.* **2012**, *11*, 174–177. [CrossRef] [PubMed]
27. Zheng, J.; Mehrvarz, H.; Ma, F.-J.; Lau, C.F.J.; Green, M.A.; Huang, S.; Ho-Baillie, A.W.Y. 21.8% Efficient Monolithic Perovskite/Homo-Junction-Silicon Tandem Solar Cell on 16 Cm 2. *ACS Energy Lett.* **2018**, *3*, 2299–2300. [CrossRef]
28. Koo, B.-K.; Lee, D.-Y.; Kim, H.-J.; Lee, W.-J.; Song, J.-S.; Kim, H.-J. Seasoning Effect of Dye-Sensitized Solar Cells with Different Counter Electrodes. *J. Electroceramics* **2006**, *17*, 79–82. [CrossRef]
29. Ramasamy, E.; Lee, W.J.; Lee, D.Y.; Song, J.S. Spray Coated Multi-Wall Carbon Nanotube Counter Electrode for Tri-Iodide (I3-) Reduction in Dye-Sensitized Solar Cells. *Electrochem. Commun.* **2008**, *10*, 1087–1089. [CrossRef]

30. Amiri, O.; Salavati-Niasari, M. High Efficiency Dye-Sensitized Solar Cells (9.3%) by Using a New Compact Layer: Decrease Series Resistance and Increase Shunt Resistance. *Mater. Lett.* **2015**, *160*, 24–27. [CrossRef]
31. Ahliha, A.H.; Nurosyid, F.; Supriyanto, A.; Kusumaningsih, T. Optical Properties of Anthocyanin Dyes on TiO_2 as Photosensitizers for Application of Dye-Sensitized Solar Cell (DSSC). *IOP Conf. Ser. Mater. Sci. Eng.* **2018**, *333*, 012018. [CrossRef]
32. Hariram Nile, S.; Hwan Kim, D.; Keum, Y. Determination of Anthocyanin Content and Antioxidant Capacity of Different Grape Varieties. *Ciência e Técnica Vitivinícola* **2015**, *30*, 60–68. [CrossRef]
33. Alighourchi, H.R.; Barzegar, M.; Sahari, M.A.; Abbasi, S. Effect of Sonication on Anthocyanins, Total Phenolic Content, and Antioxidant Capacity of Pomegranate Juices. *Int. Food Res. J.* **2013**, *20*, 1703–1709.
34. TSUDA, T. Anthocyanins as Functional Food Factors— Chemistry, Nutrition and Health Promotion. *Food Sci. Technol. Res.* **2012**, *18*, 315–324. [CrossRef]
35. Dhafina, W.A.; Salleh, H.; Daud, M.Z.; Ghazali, M.S.M. Low Cost Dye-Sensitized Solar Cells Based on Zinc Oxide and Natural Anthocyanin Dye from Ardisia Elliptica Fruits. *Optik* **2018**, *172*, 28–34. [CrossRef]
36. Jasim, K.E. Quantum Dots Solar Cells. In *Solar Cells—New Approaches and Reviews*; InTech: London, UK, 2015.

Article

Synthesis and Characterization of Titania-Coated Hollow Mesoporous Hydroxyapatite Composites for Photocatalytic Degradation of Methyl Red Dye in Water

Farishta Shafiq, Simiao Yu, Yongxin Pan and Weihong Qiao *

State Key Laboratory of Fine Chemicals, School of Chemical Engineering, Dalian University of Technology, Dalian 116024, China; farishtashafiq1@outlook.com (F.S.); ysm@mail.dlut.edu.cn (S.Y.); pyx12345106@mail.dlut.edu.cn (Y.P.)
* Correspondence: qiaoweihong@dlut.edu.cn

Abstract: Hollow mesoporous hydroxyapatite (HM-HAP) composites coated with titania are prepared to increase the stability and catalytic performance of titania for azo dyes present in the wastewater system. In this work, HM-HAP particles were first synthesized by a hydrothermal method utilizing the $CaCO_3$ core as a template and then coated with titania to form TiO_2/HM-HAP composites. Utilizing SEM, XRD, XPS, BET, FTIR, EDS, UV–vis DRS spectroscopy, and point of zero charge (PZC) analysis, the coating morphological and physicochemical parameters of the produced samples were analyzed. The photocatalytic efficiency of the synthesized coated composites was assessed by the degradation of methyl red (MR) dye in water. The results indicated that TiO_2/HM-HAP particles could efficiently photodegrade MR dye in water under UV irradiation. The 20% TiO_2/HM-HAP coating exhibited high catalytic performance, and the degradation process was followed by the pseudo-first-order (PFO) kinetic model with a rate constant of 0.033. The effect of pH on the degradation process was also evaluated, and the maximum degradation was observed at pH 6. The analysis of degraded MR dye products was investigated using LC-MS and FTIR analysis. Finally, a good support material, HM-HAP for TiO_2 coatings, which provides a large number of active adsorption sites and has catalytic degradation performance for MR dye, was revealed.

Keywords: TiO_2-HAP composites; hydrothermal synthesis; photodegradation; azo dye removal; degradation kinetics; degradation mechanism

Citation: Shafiq, F.; Yu, S.; Pan, Y.; Qiao, W. Synthesis and Characterization of Titania-Coated Hollow Mesoporous Hydroxyapatite Composites for Photocatalytic Degradation of Methyl Red Dye in Water. *Coatings* **2024**, *14*, 921. https://doi.org/10.3390/coatings14080921

Academic Editors: Joaquim Carneiro and Anton Ficai

Received: 9 June 2024
Revised: 18 July 2024
Accepted: 22 July 2024
Published: 23 July 2024

Copyright: © 2024 by the authors. Licensee MDPI, Basel, Switzerland. This article is an open access article distributed under the terms and conditions of the Creative Commons Attribution (CC BY) license (https://creativecommons.org/licenses/by/4.0/).

1. Introduction

The majority of dyes used in textiles, leather, cosmetics, food processing, ink, and paper are azo dyes, which are categorized by the existence of one or more azo groups in their chemical structure. Furthermore, 15% of the world's manufactured dyes are lost during synthesis and application as wastes that pose a threat to human health and the environment because of their toxic nature [1]. A high concentration of color, suspended particles, and COD also characterize the polluted water discharged by various industries. Even in minute quantities, the effluents of the dyes are conspicuous and hazardous. These contaminants severely impact water bodies and the plants and animals that depend on them. It is commonly known that methyl red dye is used in textile industries, paper printing, and as a pH indicator in the laboratory. However, it may create problems with the eyes, skin, and digestive tract when swallowed or inhaled [2]. Thus, owing to the health risks posed by dyes to human health, it is of the utmost importance to develop effective methods for their effective removal from wastewater.

Numerous methods have been introduced for the treatment of waste, which include chemicals such as ozonation and chlorination [3,4], physical such as adsorption [5,6] and membrane filtration [7,8], photocatalysis [9,10], and biological treatment methods [11].

Among them, photocatalytic technology provides a simple and inexpensive way to eliminate organic and inorganic contaminants from wastewater, as photocatalytic degradation may break down or mineralize the majority of organic pollutants. Among the many photocatalysts, titanium dioxide (TiO_2) is the most widely used because of its non-toxicity, low cost, and high photochemical stability, but the disadvantages of limited pollutant adsorption capacity [12–15] and significant recombination of photogenerated electron–hole pairs hinder its photocatalytic efficiency [16]. Several approaches have been implemented to improve photocatalytic performance in an effort to address these problems.

Recent research has demonstrated the capability of TiO_2-based composites to improve photocatalytic activity. For example, Sukhadeve et al. (2023) revealed the effective degradation of methylene blue dye using Ag-Fe co-doped TiO_2 nanoparticles [17]. Furthermore, Ali et al. (2023) reviewed the catalytic activity of Ag- and Zn-doped TiO_2 nano-catalysts for the removal of methylene blue and methyl orange dyes [18]. Liza et al. (2024) conducted an additional study that examined the impact of Ag-doping on the morphology, band gap, and photocatalytic activity of TiO_2 nanoparticles in the context of textile dye degradation [19]. Moreover, TiO_2 has been implemented in a variety of applications beyond wastewater treatment, including antibacterial remedies [20], protective coatings for metal prostheses [21], and self-cleaning surfaces for air purification [22].

Recent publications have indicated that TiO_2-supported materials, such as silica, hydroxyapatite (HAP), zeolite, pure natural diatomite, and activated carbon, can provide a large number of active adsorption sites, resulting in rapid mass transport and catalytic processes [23–26]. Among all these supporting materials, hydroxyapatite (HAP) has been extensively researched because of its mechanical stability, high biocompatibility, non-toxicity, and inexpensive cost [27–30]. Hydroxyapatite (HAP) is also extensively recognized for its applications in a variety of areas, such as the environmental, biomedical, and industrial sectors [31,32]. HAP's ecological friendliness is one of its most intriguing features, as it can be produced from waste materials, rendering it a sustainable and environmentally responsible option. Recent research has investigated the potential of HAP, which is derived from sources of waste, such as egg shells, fly ash, and fish bones, to reduce environmental impacts and promote sustainable practices [33–35].

Hydroxyapatite (HAP) possesses hydroxyl groups (OH) and adsorbed H_2O molecules on its surface that can interact with h^+ to create hydroxyl radicals ($^\cdot OH$), hence enhancing photocatalytic efficiency. Moreover, during the photocatalytic process, the electron phase shift of the PO_4^{3-} groups on the HAP surface can also result in the formation of $^\cdot O_2$ radicals [36]. Consequently, it is anticipated that integrating the advantages of HAP and TiO_2 will not only increase the capability of TiO_2 to absorb contaminants but also reduce the combination of photogenerated electron–hole pairs. Therefore, the use of TiO_2 in conjunction with HAP may prove to be an effective wastewater treatment method.

Prior research on TiO_2/HAP composites has mainly concentrated on using rod-shaped or quasi-spherical hydroxyapatite (HAP) structures. However, these structures often face difficulties such as particle agglomeration and ineffective TiO_2 loading [37–39]. These problems might result in a decrease in the number of active sites on the surface, which in turn reduces the effectiveness of the composite in interacting with contaminants. Our research presents a novel method that utilizes a template technique to create hollow mesoporous HAP structures, which are then coated with TiO_2. This innovative approach effectively solves the issues of aggregation that are typically seen with traditional HAP forms while also greatly improving the surface area and ease of access to the active areas inside the composite. By employing hollow mesoporous hydroxyapatite (HAP) in this distinct morphology, the effectiveness and durability of the TiO_2/HAP composite for photocatalytic and environmental purposes are significantly enhanced. This study is the first known case of employing hollow mesoporous HAP templates for TiO_2 coatings. This opens new prospects for improving the efficiency and adaptability of composite materials in sustainable technologies. The unique structure of spherical hollow hydroxyapatite particles not only prevents aggregation but also allows for a greater surface area owing to the hollow interior. This makes them very appropriate for coating TiO$_2$ and conducting investigations on photocatalysis.

In this research work, we prepared HM-HAP particles through a hydrothermal technique using the CaCO$_3$ core as a template. The synthesized HM-HAP was then coated with different amounts of TiO$_2$ to form a TiO$_2$-coated HM-HAP composite. Figure 1 shows a schematic diagram of the TiO$_2$/HM-HAP particle synthesis. The synthesized TiO$_2$/HM-HAP composites exhibited good photocatalytic activity for MR dye removal under UV light. XRD, XPS, BET, SEM, and UV–vis DRS were used to examine the effect of TiO$_2$ coatings on the structural behavior of HM-HAP. The photocatalytic degradation of methyl red (MR) dye under UV irradiation was investigated over the TiO$_2$/HM-HAP composites. Hence, this research paper presents results on a good support material, HM-HAP, for TiO$_2$ coatings, which provides a large number of adsorption sites, resulting in rapid mass transport and catalytic processes, and finally studies the photocatalytic degradation of MR dye.

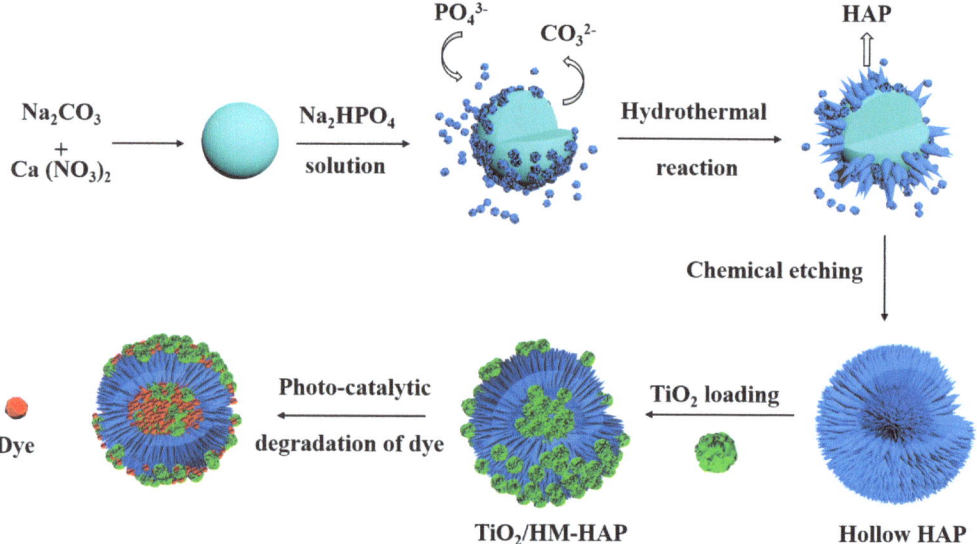

Figure 1. Schematic diagram illustrating the synthesis of TiO$_2$/HM-HAP particles.

2. Experimental Methods

2.1. Materials

Sodium carbonate (Na$_2$CO$_3$), calcium nitrate (Ca(NO$_3$)$_2$), disodium hydrogen phosphate dodecahydrate (Na$_2$HPO$_4$·12H$_2$O), potassium hydrogen phthalate (KHP), potassium chloride (KCl), potassium dihydrogen phosphate (KH$_2$PO$_4$), sodium hydroxide (NaOH), and hydrochloric acid (HCl) were purchased from DAMAO. Sodium poly (styrene sulfonate) (PSS, Mw = 70,000), methyl red (MR), and Titanium dioxide (TiO$_2$) were purchased from MACKLIN.

2.2. Synthesis of HM-HAP Particles

A 0.1 M solution of Na$_2$CO$_3$ (20 mL) was quickly mixed with a 0.1 M Ca (NO$_3$)$_2$ solution (20 mL) containing 150 mg of PSS under magnetic stirring (600 rpm). The solution temperature was maintained at 30 °C for about 30 min. The obtained CaCO$_3$ particles, resulting from the quick precipitation procedure, were rinsed three times with deionized water before being collected by centrifugation (7000 rpm, 5 min). The obtained CaCO$_3$ particles were then added to a Na$_2$HPO$_4$ solution (30 mL, 0.8 M). The resultant mixture was then placed in an autoclave of 100 mL capacity and heated at 140 °C for 4 h. After the reaction was complete, the $CaCO_3$ core was removed by adding a few drops of acid to the solution and stirring it for 2 h.

2.3. Synthesis of TiO_2/HM-HAP Composite

The obtained materials, HM-HAP (150 mg) and titanium dioxide (TiO_2), were mixed in various amounts with butanol and agitated for one hour at 45 °C. The butanol was then evaporated at 60 °C using a rotary evaporator. The sample was then heated for 4 h at 450 °C. The TiO_2 was coated onto the HM-HAP at 10%, 20%, 30%, 40%, and 50%.

2.4. Characterization of TiO_2/HM-HAP Composite

The material's surface shape and size were determined using a scanning electron microscope HITACHI-SU8220 (High-Tech corporation, Tokyo, Japan). The X-ray diffraction (XRD) patterns were acquired using a smart lab X-ray diffractometer (XRD, Rigaku corporation, Tokyo, Japan) with Cu Kα (λ = 0.15406 nm) radiation. On a ThermoFisher—6700 (Waltham, MA, USA) FTIR spectrometer, the material's FTIR spectrum was obtained using the KBr pellet technique. The pore structure and specific surface area of the coated particles were determined using a BET surface area analyzer (BSD Instrument Technology (Beijing, China) Co., Ltd. (BSD-PS (M)) at 77 K, utilizing helium as the carrier gas at liquid nitrogen temperature. The formation of TiO_2 coatings on the HM-HAP particles was confirmed via an energy-dispersive X-ray spectrometer-QUANTA 450 (Thermo Fisher Scientific, Waltham, MA, USA). X-ray photoelectron spectroscopy (XPS) data were collected on a Thermo Scientific photoelectron spectrometer—ESCALAB250Xi (Thermo Fisher Scientific, Winsford, UK). UV-vis diffuse reflectance spectra (DRS) of the synthesized materials were measured by a UV-vis spectrophotometer (Lambda 1050+, Perkin Elmer, Shanghai, China) utilizing $BaSO_4$ as a reference material. A spectroscopic study utilizing a UV-vis spectrophotometer carry-100 (Agilent, Kuala Lumpur, Malaysia) was used to evaluate the absorbance of methyl red dye throughout its photocatalytic degradation. The absorbance spectra were recorded between 350 and 750 nm wavelength. The pH measurement was performed using a PHS-3C pH meter (Shanghai Yoke Instrument Co., Ltd., Shanghai, China) with an integrated temperature sensor for temperature regulation. The examination of degradation products was conducted using liquid chromatography–mass spectrometry (LC-MS, G6230, Agilent, Santa Clara, CA, USA). The TOC of before and after decolorization of the dye solution was quantified by an organic element analyzer (UNICUBE, Elementar Analysensysteme GmbH, Langenselbold, Germany).

2.5. Point of Zero Charge (PZC)

Measuring PZC aids in predicting the ionic character of a catalyst, which clarifies the interaction process between a dye and the catalyst. The salt addition method was used to measure the PZC point of the formed TiO_2-coated HM-HAP composites. Typically, a 0.1 M NaCl solution was prepared, and for each experiment, 50 mL of the solution was taken in a 100 mL beaker. To each beaker, 30 mg of TiO_2-coated HM-HAP was added and then stirred for about 3 h. Each solution's pH was maintained using a mixture of buffers (pH 2 to 12).

2.6. Photocatalytic Studies of MR

The photocatalytic activity of the samples was evaluated by observing the degradation of MR dye as the pollutant target under ultraviolet irradiation. The chemical structure and other characteristics of the employed dye are detailed in Table S1. Initially, 25 mL of MR solution (10 mg/L) and 30 mg of catalyst were vigorously agitated for 30 min in the dark to produce an adsorption/desorption equilibrium. The solution was then agitated at room temperature under UV light irradiation (λ = 254 nm) for several minutes. Periodically, 3 mL of the solution was collected and filtered to eliminate the solid phase. Although this decreased the initial volume, parallel experiments were carried out to ensure uniformity and eliminate potential mistakes resulting from volume reduction. More precisely, 25 mL samples were prepared and subjected to the same circumstances. This enabled the extraction of small portions from various samples at each specific time interval. This method guaranteed that the decrease in volume did not impact the accuracy of the outcomes. After that, the absorbance of the filtrates was measured spectroscopically at 437 nm for MR dye. In order to measure the extent of degradation of the MR dye, a calibration curve was employed, which can be seen in Figure S1 in the Supplementary Information. The calibration curve was generated by measuring the absorbance of a range of MR dye solutions with predetermined concentrations, enabling us to establish a precise relationship between absorbance and dye concentration. Applying this calibration curve, the concentration of MR dye at various irradiation times was established, facilitating the calculation of the photodegradation rate and efficiency.

Photolysis and adsorption experiments were also conducted under the same conditions as the photocatalytic experiment.

The effect of various parameters, such as pH and contact time, on the degradation of MR by the catalysts was also analyzed. The solution pH (2, 4, 6, 8, and 10) was monitored by the addition of sodium hydroxide (NaOH), sodium bicarbonate ($NaHCO_3$), potassium hydrogen phthalate (KHP), potassium chloride (KCl), hydrochloric acid (HCl), and potassium dihydrogen phosphate (KH_2PO_4).

The percentage degradation of the MR dye was determined using the following relationship [40]:

$$\% \ Degradation = \frac{[MR]_0 - [MR]_t}{[MR]_0} \times 100 \tag{1}$$

where $[MR]_0$ represents the initial concentration and $[MR]_t$ represents the concentration at time t of the dye MR.

2.7. Kinetics of MR Degradation

The obtained experimental data were assessed further using the (PFO) pseudo-first-order kinetic model, which is represented by the given expression [41]:

$$ln\frac{[MR]_0}{[MR]_t} = k_1 t \tag{2}$$

where t is the given time, $[MR]_0$ symbolizes the initial concentration, and $[MR]_t$ symbolizes the concentration at any time (t) of MR dye. k_1 is the rate constant, and its values can be derived from the slope of the graph $ln[MR]_0/[MR]_t$ vs. reaction time.

3. Results and Discussion

3.1. Synthesis of TiO_2/HM-HAP Composite

In this research study, TiO_2-coated HM-HAP composites were produced and employed in a photocatalytic study. First, HM-HAP particles were synthesized using a hydrothermal method with PSS-doped vaterite $CaCO_3$ as a hard template. The presence of PSS, a polyelectrolyte with a substantial negative charge, produces negatively charged HM-HAP particles. In addition, PSS (sodium poly (styrene sulfonate)) was used as a crystal growth additive to accelerate the transformation of $CaCO_3$ from calcite to vaterite during the $CaCO_3$ manufacturing process. The reagents Na_2CO_3 and Ca $(NO_3)_2$ were utilized first for the synthesis of $CaCO_3$ by the formation of precipitates. The obtained $CaCO_3$ particles were then added to Na_2HPO_4, and after the completion of the reaction, the core was removed by an etching process carried out with acetic acid. The synthesized HM-HAP particles were then coated with a range of TiO_2 concentrations (10% to 50%) to create a TiO_2-coated HM-HAP composite. FTIR and XRD analysis confirmed the synthesis of HM-HAP and TiO_2-coated HM-HAP composites.

The FTIR spectra of pure TiO_2, HM-HAP, and TiO_2-coated HM-HAP composites are shown in Figure 2a. The functional peaks at 3432 cm^{-1} and 1632 cm^{-1} are the results of lattice water in the samples [42]. At 634 cm^{-1}, the distinctive broad absorption bands of TiO_2 are detected, which can be ascribed to the stretching vibration of Ti–O–Ti bonds [43]. The distinctive bands at 563 and 602 cm^{-1} for the HM-HAP sample correspond to the bending vibration of PO_4^{3-}, whereas the band at 1031 cm^{-1} corresponds to the stretching vibration of PO_4^{3-}. The characteristic CO_3^{2-} bands are situated at 1466, 1410, and 874 cm^{-1} [44–46]. The spectra of TiO_2-coated HM-HAP (10%–50%) composites show that the characteristic peaks of TiO_2 and HM-HAP are well preserved, indicating that the structure transformation of TiO_2 and HM-HAP did not change during composite production. This demonstrates the effective synthesis of TiO_2-coated HM-HAP composites, which is compatible with the XRD data.

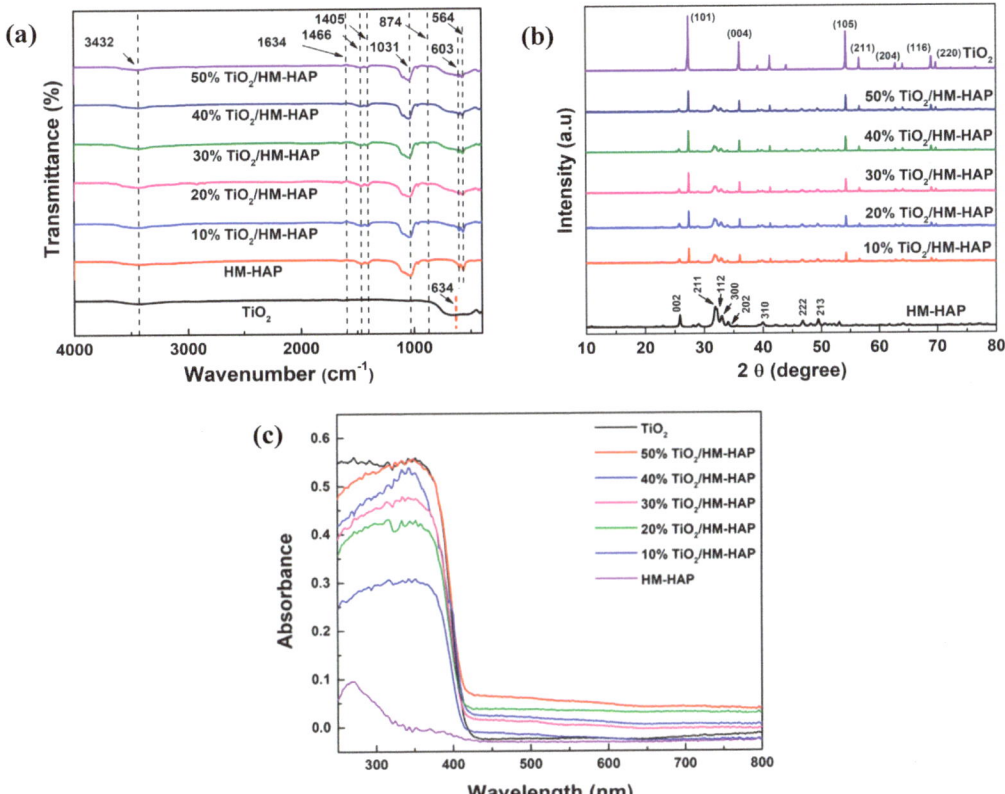

Figure 2. FTIR spectra of as-synthesized samples (**a**), XRD spectra of TiO_2, HM-HAP, and TiO_2-coated HM-HAP (10%, 20%, 30%, 40%, and 50%) composites (**b**), and UV–vis DRS spectra of pure TiO_2, HM-HAP, and TiO_2-coated HM-HAP composites (**c**).

The XRD patterns of TiO_2, HM-HAP, and TiO_2-coated HM-HAP composites are shown in Figure 2b. The diffraction peaks at 25.2°, 37.1°, 53.9°, 55.8°, 62.7°, 68.5°, and 70.3° correspond to the (101), (004), (105), (211), (204), (116), and (220) planes of anatase-TiO_2, respectively (JCPDS no. 21-1272) [47]. In the case of pure HM-HAP, all diffraction peaks and their relative intensities correspond to the standard diffraction data of pure hexagonal phase HAP (JCPDS, 09-0432) [48,49]. Specifically, the peaks at two values of 31.9°, 32.1°, and 33.1° correspond to the (211), (112), and (300) planes of HAP, respectively. These indices confirm the successful synthesis of hydroxyapatite with a well-defined crystalline structure. In addition, the diffraction patterns of the TiO_2-coated HM-HAP composites (10%–50%) demonstrate the existence of both anatase TiO_2 and hexagonal phase HAP. Thus, the TiO$_2$-coated HM-HAP composites were successfully synthesized using the hydrothermal technique.

3.2. Characterization of the TiO$_2$/HM-HAP Composite

After confirming that the material was synthesized, we further characterized a series of samples with different Ti contents. Figure 2c depicts the UV–vis DRS spectra of pure TiO$_2$, HM-HAP, and TiO$_2$-coated HM-HAP composites. The band gap energy values for the as-synthesized samples are illustrated in Table S2, and their respective graphs are shown in Figure S2. It can be seen that the absorption edge of pure TiO$_2$ is evidently located around 429 nm, whereas the absorption edge of pure HM-HAP is less than 350 nm. Compared with pure HM-HAP, the TiO$_2$-coated HM-HAP composites exhibit a considerable red shift in the absorption edge. In addition, the absorption edge of the TiO$_2$-coated HM-HAP composites exhibit a little blue shift relative to that of TiO$_2$, indicating that the TiO$_2$-coated HM-HAP composites have a greater oxidation capability than TiO$_2$.

The size and morphology of HM-HAP and TiO$_2$-coated HM-HAP composites were visualized using SEM (Figure 3a–f). The morphology of the particles revealed a spherical shape with a hollow interior. The micrographs indicated that the particle size of HM-HAP was around 1–2 µm. All six formulations of HM-HAP had the same morphology and shape.

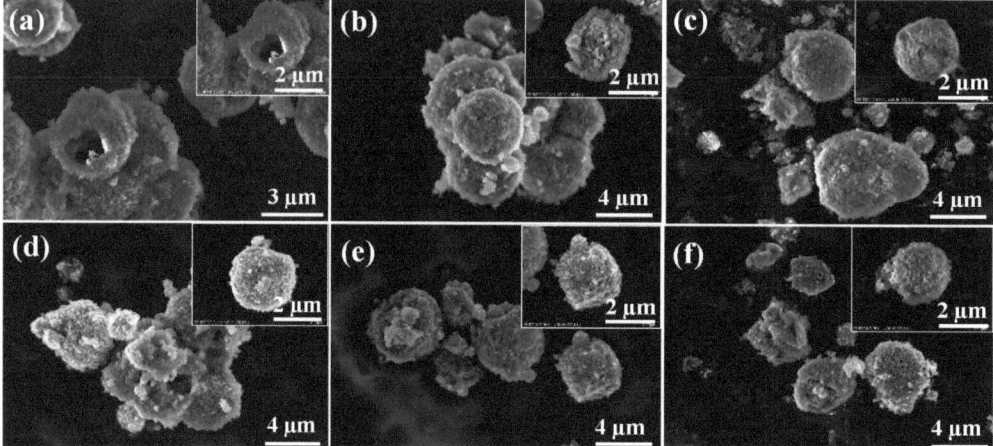

Figure 3. SEM micrographs of (**a**) HM-HAP and (**b**) 10% TiO$_2$/HM-HAP, (**c**) 20% TiO$_2$/HM-HAP, (**d**) 30% TiO$_2$/HM-HAP, (**e**) 40% TiO$_2$/HM-HAP, and (**f**) 50% TiO$_2$/HM-HAP coated composites. In (**a–f**), the insets depict the high magnifications corresponding to those figures.

Energy-dispersive X-ray spectroscopy (EDS) was used to examine the surface elemental analysis of TiO$_2$-coated HM-HAP composites. The levels of three elements, i.e., Ca, P, and Ti, in the samples, are summarized in Figure 4. The mapping profile illustrates the homogenous surface distribution of the three components. All TiO$_2$-coated HM-HAP composites exhibited a close interaction between these elements. The corresponding EDS spectra in Figure 5a confirmed the existence of the Ti element coming from TiO$_2$, demonstrating the successful formation of a TiO$_2$-coated HM-HAP composite. Hence, the incorporation of TiO$_2$ into HM-HAP particles may therefore be validated.

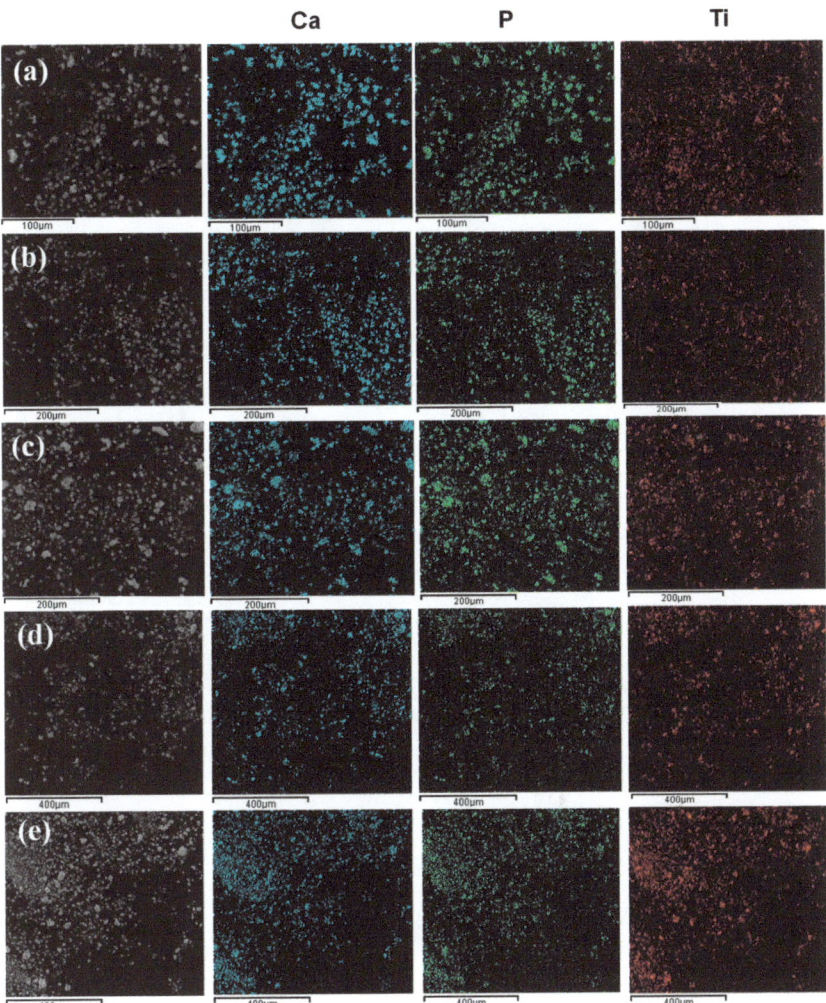

Figure 4. EDS elemental mapping of (**a**) 10% TiO$_2$/HM-HAP, (**b**) 20% TiO$_2$/HM-HAP, (**c**) 30% TiO$_2$/HM-HAP, (**d**) 40% TiO$_2$/HM-HAP, and (**e**) 50% TiO$_2$/HM-HAP coated composites.

The pore volume, pore size, and Brunauer–Emmett–Teller (BET) specific surface area acquired from N$_2$ adsorption–desorption for all the TiO$_2$-coated HM-HAP composites are presented in Table 1. Figure 5b depicts the isotherms and corresponding pore size distribution histograms, indicating the mesoporous structure of the TiO$_2$-coated HM-HAP composites. According to the IUPAC system, all the synthesized samples have type IV isotherms, which have an H3 hysteresis loop. This is because particle aggregation makes slit-shaped pores [50]. The BET-specific surface area of 10% TiO$_2$-coated HM-HAP was 56 m^2/g and declined as the TiO$_2$ concentration increased, while the corresponding pore diameter increased gradually. A possible explanation for this result is that when the number of TiO$_2$ molecules increases, aggregation may occur, resulting in no greater pore occupancy. Thus, a decreasing trend in the specific surface area occurred. Moreover, the coating of TiO$_2$ onto HAP greatly increases the surface area of the resulting composite material (Table 1). The augmentation in surface area is essential as it offers a greater number of active sites for photocatalytic reactions, hence enhancing the efficacy of the photodegradation process. More precisely, the TiO$_2$-coated HM-HAP composite has a greater surface area in comparison with pure HAP, as specified in Table 1. The improvement can be attributed to the synergistic interplay between TiO_2 and HAP, with HAP serving as a supporting framework that hinders the clumping of TiO_2 nanoparticles, hence ensuring a large surface area.

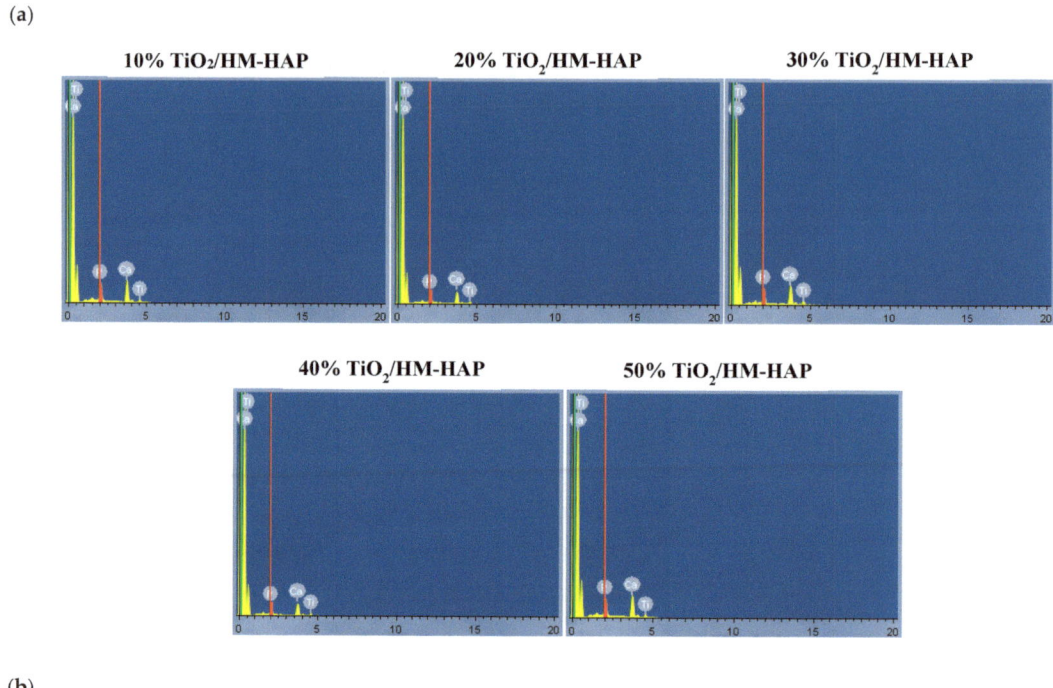

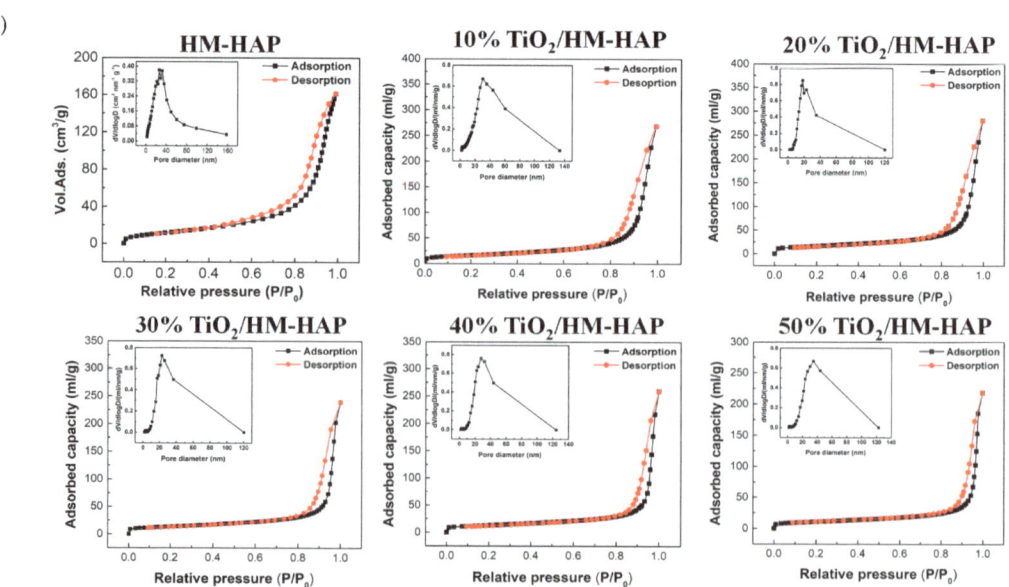

Figure 5. EDS spectra (**a**) and N_2 adsorption–desorption isotherms and pore size distribution (**b**) of the synthesized TiO_2/HM-HAP coated composites.

Table 1. BET-specific surface area and pore diameter analysis of the synthesized composites.

Sample	BET Specific Surface Area (m^2/g)	Pore Size (nm)
HM-HAP	44	12.47
10% TiO$_2$/HM-HAP	56	27.82
20% TiO$_2$/HM-HAP	57	28.19
30% TiO$_2$/HM-HAP	47	29.28
40% TiO$_2$/HM-HAP	46	31.94
50% TiO$_2$/HM-HAP	38	33.43

The surface chemical composition and chemical states of pure TiO$_2$, HM-HAP, and TiO$_2$-coated HM-HAP composites were investigated using XPS. The XPS survey spectra of TiO$_2$, HM-HAP, and 20% TiO$_2$/HM-HAP coatings are depicted in Figure 6. In Figure 6a, the survey spectra of TiO$_2$-coated HM-HAP reveal the presence of the four elements including Ti, Ca, P, and O. As indicated in the figure, the additional C element peak is mostly produced from adventitious carbon. The XPS peaks in the Ti 2p of TiO$_2$ in Figure 6b, situated at a binding energy of 458.6 eV and 464.2 eV, are ascribed to the Ti-O bonds [51] but in the 20% TiO$_2$/HM-HAP coating, the Ti 2p peaks shift to lower energy levels, 458.1 eV and 463.9 eV, respectively, which may be due to the presence of Ti-O-Ca bonds in the TiO$_2$/HM-HAP coated composite. Likewise, the Ca 2p peaks of HAP shown in Figure 6c are situated at 346.8 eV and 350.4 eV, whereas in the 20% TiO$_2$/HAP coating, the Ca 2p peaks slightly shift towards the higher binding energy, i.e., 347.1 eV and 350.7 eV [52]. Figure 6d displays the high-resolution O 1s spectra that correspond to TiO$_2$, HM-HAP, and TiO$_2$/HM-HAP coated composites, respectively. Ti-O bonds and -OH groups are responsible for the two peaks in TiO$_2$ that are located at 528.9 and 530.4 eV, respectively. In the HM-HAP spectra, the phosphate group (PO$_4^{3-}$) and adsorbed water are responsible for the two peaks that are positioned at 530.8 eV and 532.1 eV, respectively [53,54]. In the case of the 20% TiO$_2$/HM-HAP coated composite, the three peaks at 529.1 eV, 529.4 eV, and 531.1 eV are ascribed to the lattice oxygen species Ti-O bonds (TiO$_2$), PO$_4^{3-}$ (phosphate group), and O-C bond of the CO$_3^{-2}$, respectively [54]. Hence, the analysis indicated that TiO$_2$ particles existed in the formed composition.

Moreover, the charge transfer between HAP and TiO$_2$ was also confirmed by the XPS analysis of the samples. By comparing the XPS data of individual atoms in pure TiO$_2$ and HM-HAP with those in the composite material, significant shifts in binding energy peaks were observed. These shifts in binding energy are compelling evidence of alterations in the chemical environment, suggesting the occurrence of charge transfer between HAP and TiO$_2$. Hence, the shift in peaks not only signifies the formation of the composite material but also provides clear evidence of the charge transfer process taking place between the two components.

Furthermore, the stability of the synthesized photocatalyst can be supported by the fact that hydroxyapatite (HAP) exhibits exceptional stability up to 1000 °C, as supported by rigorous testing through Fourier Transform Infrared (FTIR) analysis and Thermogravimetric Analysis (TGA) [55,56]. Therefore, in the formation of the TiO$_2$ composite with HAP, the inherent stability of HAP was utilized, which inherently imparts durability to the resulting photocatalyst. This stability is not merely asserted but substantiated through meticulous examination, as evidenced in various papers. The FTIR and TGA analysis presented in the literature elucidates the stability of HAP at different temperatures, thereby confirming the stable nature of the synthesized photocatalyst.

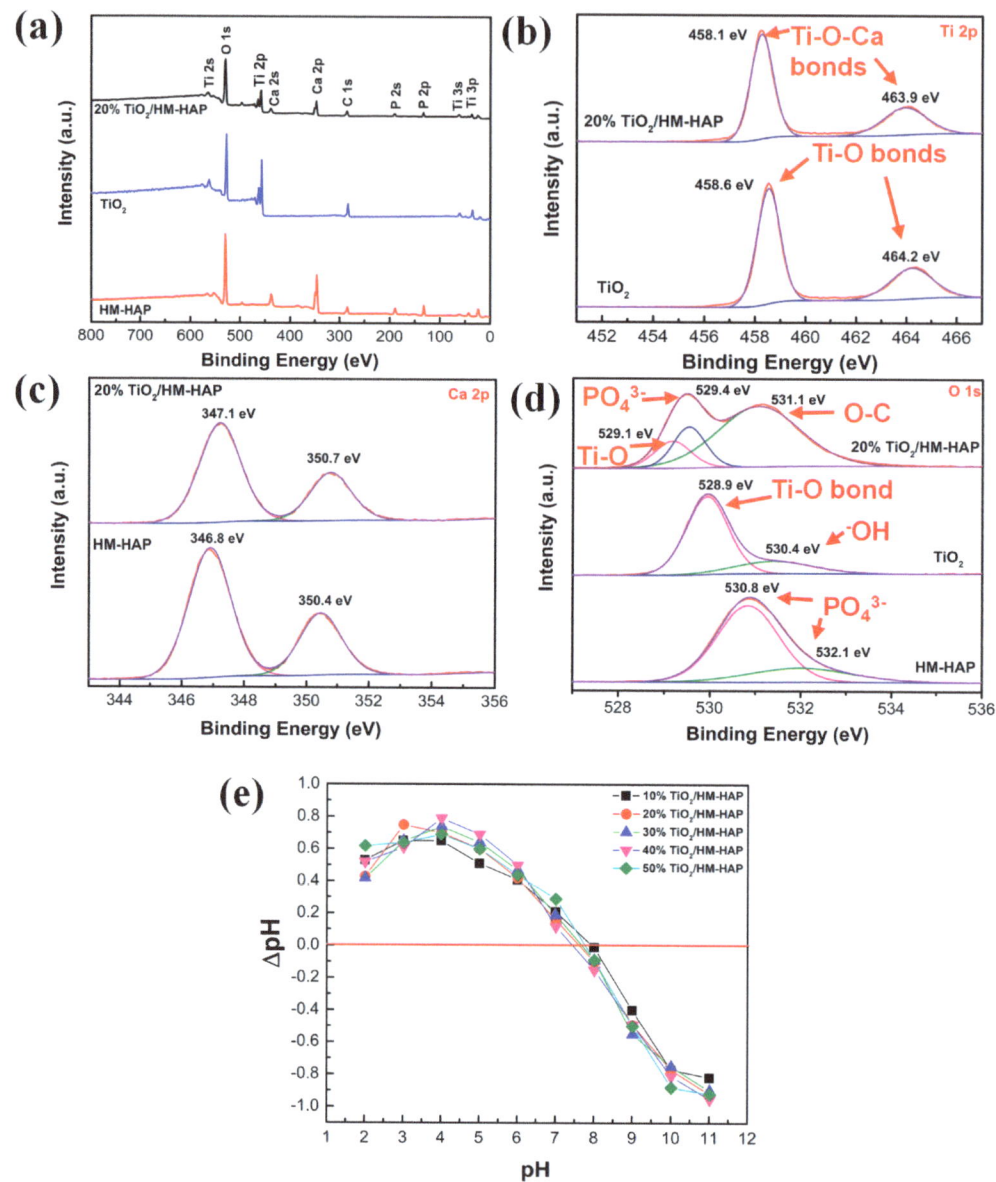

Figure 6. XPS survey spectra (**a**) and high-resolution XPS spectra of Ti 2p (**b**), Ca 2p, (**c**), O 1s (**d**), and PZC plot (**e**) of the synthesized samples.

PZC is regarded as one of the several methods for determining the kind of surface charge. It is the value at which the surface charge density of a material is equal to zero. Consequently, the PZC points of the TiO_2/HM-HAP coated composites were determined. According to Figure 6e, the PZC point of the TiO_2/HM-HAP composites (10%–50%) is between 7.5 and 8.1. The detailed PZC values of each composite are shown in Table S3. This signifies that the synthesized composites have a negative charge above this number and a positive charge below it.

In conclusion, the findings of several characterizations revealed that the TiO_2/HM-HAP coated composite was successfully formed. The SEM micrographs indicated the spherical morphology of the particles having mesopores, as confirmed by the analysis of N_2 adsorption–desorption. The EDS elemental mapping and the corresponding spectra clearly showed the existence of Ti in the composite. Furthermore, the XPS analysis validated the peak corresponding to the lattice oxygen species of TiO_2 in the composite of TiO_2-coated HM-HAP.

3.3. Evaluation of the Photocatalytic Performance of the Synthesized Composites for the Degradation of MR

The photocatalytic activity of the synthesized TiO_2-coated composites was assessed by the photodegradation of MR under UV light irradiation. First, the linearity of MR's spectroscopic response was measured, and a graph of absorbance against concentration was plotted to demonstrate the conformity of the dye to the Lambert–Beer law (Figure S1). As illustrated in Figure 7a, when catalyzed by a TiO_2-coated HM-HAP composite with UV irradiation (60 min), the degradation efficiency increased significantly, and in the presence of the 20% TiO_2/HM-HAP coating, up to almost 88% degradation efficiency was achieved. In contrast, pure HM-HAP and TiO_2 demonstrated degradation efficiencies of 20% and 42%, respectively (Figure 7a). This enhanced performance can be attributed to the unique properties of HAP, specifically its hydroxyl groups (^-OH) and adsorbed H_2O molecules on the surface. These hydroxyl groups are instrumental in interacting with h^+ to generate hydroxyl radicals ($^•OH$), a highly reactive species known for its effectiveness in photocatalysis. Furthermore, the photocatalytic process involves intriguing electron phase shifts within the PO_4^{3-} groups on the HAP surface. This phenomenon leads to the formation of $^•O_2$ radicals, further augmenting the catalytic efficiency of the composite material [57]. Although bare TiO_2 exhibits photocatalytic activity, its efficiency is restricted by the rapid recombination of photogenerated electron–hole pairs. Conversely, HM-HAP alone offers adsorption sites but does not possess the catalytic efficiency of TiO_2. The TiO_2/HM-HAP composite optimizes charge separation and increases the number of active sites for the degradation reaction by integrating the benefits of both materials. Hence, integrating the advantages of HAP and TiO_2 not only results in an enhancement in TiO_2's capability to absorb contaminants but also mitigates the recombination of photogenerated electron–hole pairs. This synergy between HAP and TiO_2 not only improves the material's adsorption capacity but also amplifies its overall photocatalytic performance.

However, it can also be seen that as the concentration of TiO_2 increased by over 30%, the degradation efficiency declined gradually, demonstrating a decrease in activity. Generally, adsorption and degradation capacities should increase as the TiO_2 level rises. However, this was not the situation observed, as HM-HAP with a higher TiO_2 concentration exhibited only a little increase in adsorption in comparison with the one with a lower TiO_2 concentration. The existence of a large number of catalysts can prevent light from penetrating pores and promote intermolecular collision, thereby reducing photodegradation. In addition, the large number of TiO_2 particles could have covered the active sites and inhibited the generation of the active compounds [58]. These disadvantages counteract the enhanced photocatalysis provided by additional TiO_2.

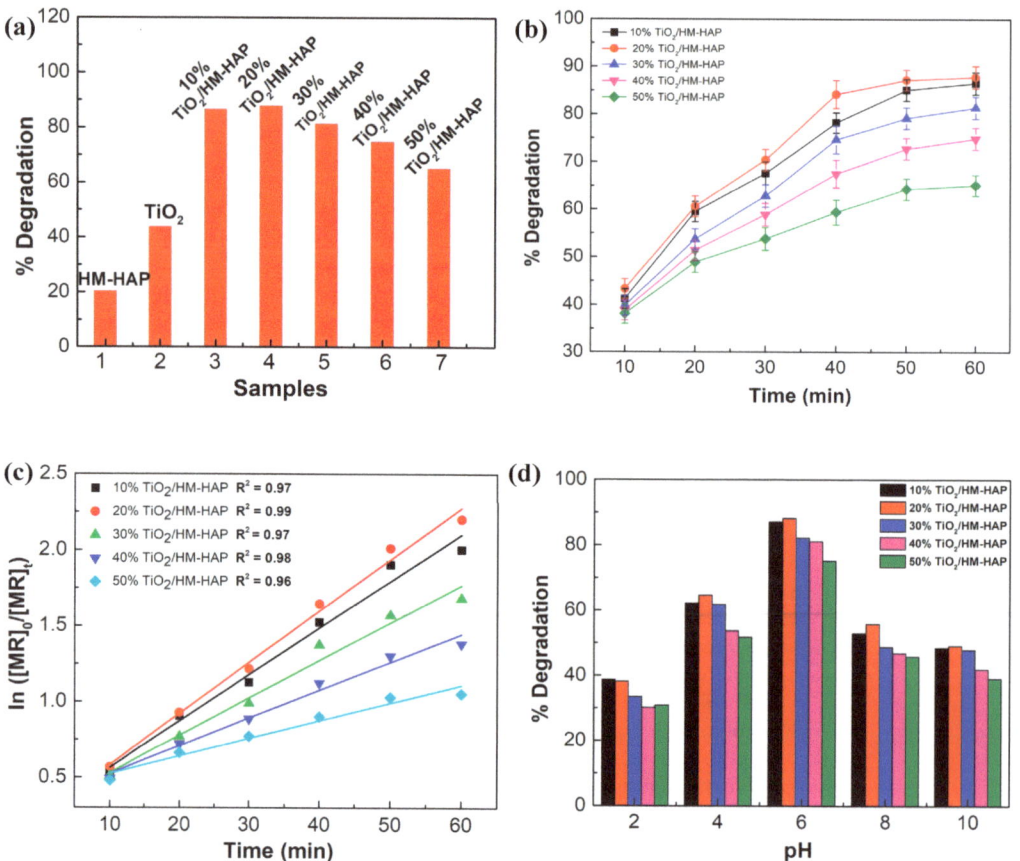

Figure 7. Degradation efficiency of MR over the synthesized composites at pH = 6, 25 °C, catalyst = 30 mg, and 60 min of irradiation time (**a**), time curve of % degradation at pH = 6, and 25 °C temperature (**b**), degradation kinetics of MR at pH = 6, and 25 °C temperature (**c**), and effect of pH on the % degradation of MR at 25 °C, and 60 min of irradiation time (**d**).

3.3.1. Photocatalytic Degradation Kinetics

Figure 7b shows the MR percent degradation versus irradiation time graphs of the synthesized TiO$_2$/HM-HAP coated composites, demonstrating the gradual decomposition of MR. Figure S3 shows the UV spectra of MR at different time intervals. The rates of MR photodegradation were evaluated using a pseudo-first-order kinetic model in order to establish quantitative comparisons. In Figure 7c, it is evident that the pseudo-first-order kinetic model suited the experimental data well, and the R^2 values were all above 0.90, indicating that the reactions were compatible with the first-order kinetic. Table 2 displays the values of the rate constant (k_{MR}) and regression coefficient (R^2) for the TiO$_2$/HM-HAP coated composites for MR removal. Among the samples, the composite containing the 20% TiO$_2$/HM-HAP coating exhibited the highest degradation rate constant, i.e., 3.3×10^{-2} min^{-1}. The k_{MR} followed the order of 20% TiO$_2$/HM-HAP > 10%, TiO$_2$/HM-HAP > 30% TiO$_2$/HM-HAP > 40% TiO$_2$/HM-HAP > 50% TiO$_2$/HM-HAP. As previously indicated, this could be because composites with low amounts of TiO$_2$ have many active sites on their surface.

Table 2. The rate constant values of the photocatalytic degradation of MR.

Sample	Rate Constant k_1 (min^{-1})	R^2
10% TiO$_2$/HM-HAP	0.030	0.97
20% TiO$_2$/HM-HAP	0.033	0.99
30% TiO$_2$/HM-HAP	0.024	0.97
40% TiO$_2$/HM-HAP	0.018	0.98
50% TiO$_2$/HM-HAP	0.011	0.96

These above observations clearly indicate that 10% TiO$_2$/HM-HAP has remarkable photocatalytic efficiency, even though it has a lower TiO$_2$ concentration compared with 20% TiO$_2$/HM-HAP. This finding indicates that 10% TiO$_2$/HM-HAP has high efficiency as a photocatalyst, attaining a degradation degree (DD) that is almost equivalent to 20% TiO$_2$/HM-HAP while containing half the amount of TiO$_2$. When comparing 10% TiO$_2$/HM-HAP to pure TiO$_2$, it is evident that 10% TiO$_2$/HM-HAP exhibits a substantially greater efficiency per unit of TiO$_2$. Furthermore, the degradation degree (DD) per unit of TiO$_2$ was calculated for each sample and is presented in Supplementary Information Section S1.

The efficiencies of different synthesized composites were compared to identify the most effective and cost-efficient photocatalyst. According to the findings, 20% TiO$_2$/HM-HAP exhibits the highest overall degrading efficiency. However, 10% TiO$_2$/HM-HAP provides a comparable performance with less TiO$_2$. The lower TiO$_2$ content of 10% TiO$_2$/HM-HAP is notably advantageous from a cost and application perspective, as it reduces material costs and increases the potential for large-scale applications.

3.3.2. Effect of pH on the Degradation of MR

The pH of a solution is a crucial variable in the photocatalytic degradation process. The pH affects both the properties of a dye (hydrophobicity, speciation behavior, and water solubility) and the surface charge of a catalyst. Below pH 5.3 (pKa), MR is mostly cationic (i.e., protonated), and above pH 5.3, it is anionic (i.e., unprotonated). It was previously stated that the electrostatic interaction among the material surface, solvent molecules, and dye during photocatalytic degradation is pH-dependent [59]. The influence of solution pH was studied to find the optimal pH range for maximal photocatalytic degradation. While examining the pH range (2, 4, 6, 8, and 10), a blue shift in λmax for MR was observed when the pH of the solution was changed from 4 to 6. No other shifts in the wavelength were observed when the pH was increased further. The UV spectra of MR dye at different pHs are presented in Figure S4.

Figure 7d shows the photocatalytic behavior of the synthesized composite towards the degradation of a solution of MR dye at different pH levels carried at room temperature. In Figure 7d, it is clear that the maximum photocatalytic degradation happened at pH 6. The explanation for this behavior is that the PZC of the TiO$_2$/HM-HAP coated composites is in the range of 7.5–8.1. Thus, the material's surface is positively charged when the pH is <7.5 and negatively charged when the pH is >8.1. Therefore, the dye was in anionic form, and the surface of the material at pH 6 was positively charged [46,60,61]. Hence, in this condition, the interaction between the two species preferred photocatalytic degradation.

3.3.3. Post-Photodegradation FTIR Analysis and Mineralization Study of MR Dye

The FTIR spectrum of the MR dye is depicted in Figure 8a. The spectrum exhibited prominent peaks at 2919 and 2857 cm^{-1}, which correspond to the stretching vibrations of the C–H for –CH$_3$ groups. The O-H stretch is represented by the band at 3430 cm^{-1}. The apparent band at 1716 cm^{-1} is the result of the C=O stretching of carbonyl groups. The C–C stretching vibration and N=N vibration of benzene are observed at 1602 cm^{-1} and 1528 cm^{-1}, respectively. At 1368 and 1152 cm^{-1}, strong bands are ascribed to C–N stretching vibrations. The minor bands detected at 1115, 818, and 764 cm^{-1} are indicative of C–H stretching vibrations [62,63].

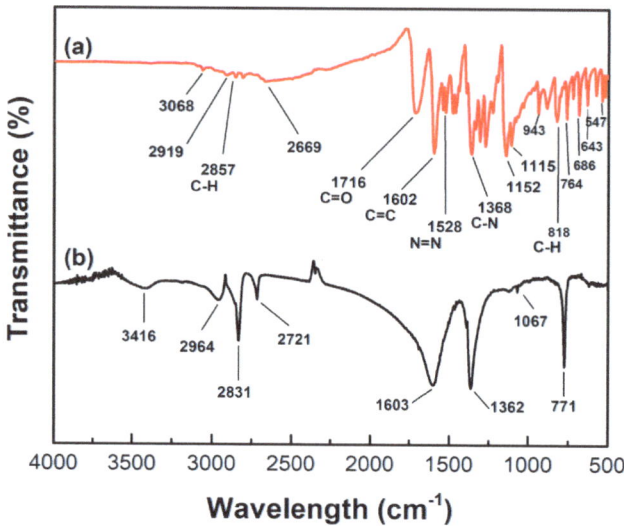

Figure 8. FTIR spectra of MR dye before degradation (**a**) and after degradation (**b**).

The FTIR analysis of the product obtained subsequent to the degradation of MR is illustrated in Figure 8b. The disappearance of the peak at 1528 cm^{-1}, which was caused by the stretching vibration of N=N, indicates the cleavage of the azo group. The disappearance of the band at 1159 cm^{-1} signifies a breakdown of C–N stretching vibrations [63,64]. The peak observed at 1067 cm^{-1} is the result of stretching variations in the C–O bond. Significant variations can be observed by comparing the FTIR spectra of the original dye with those of the degraded dye or its metabolites. Certain methyl red dye peaks were observed to have vanished after degradation, while others re-emerged subsequent to degradation; this observation suggested that the dye had transformed into new compounds or metabolites. FTIR analysis confirmed and contributed to the reduction and elimination of the methyl red dye's azo linkage. Moreover, an LC-MS [65] analysis of decolorized MR solution was also carried out. The resultant spectra are provided in Figure S5, and the identified degraded products of MR are shown in Figure S6.

As per the results of FTIR, the degradation of MR dye was confirmed. Additionally, the mineralization of MR dye was assessed by monitoring the reduction in total organic carbon (TOC) content following the treatment with the TiO$_2$/HM-HAP coated composite. At a pH of 6, a TOC removal of 35.64% was observed for 10 mg/L MR using the TiO$_2$/HM-HAP coated composite following a 1 h reaction time. Consequently, the TiO$_2$/HM-HAP coated composite was capable of facilitating dye decolorization and mineralization.

Moreover, the results of the photolysis experiment (Figure 9) demonstrated that there was minimal degradation of the MR dye when exposed just to direct UV light. This suggests that direct photolysis has a small impact on the overall degradation process. Furthermore, the findings from the adsorption experiment, presented in Figure 9, demonstrate that the catalyst was responsible for 20% of the overall removal of MR by adsorption. This demonstrates that although adsorption plays a role in the overall process, the main mechanism for removing dye under UV irradiation is definitely photodegradation. HAP possesses a significant number of hydroxyl groups, which contribute to its surface having a primarily negative charge. Methyl red (MR) is a negatively charged dye, and because of the like charges, there is a repulsive force that decreases its ability to stick to the HAP surface. While HAP does have some sites with a positive charge due to calcium ions, the overall negative charge from the many hydroxyl groups is more prominent. The structure of HAP and an explanation of the interaction between HAP and TiO$_2$ in TiO$_2$/HM-HAP composites are further presented in Supplementary Information Section S2.

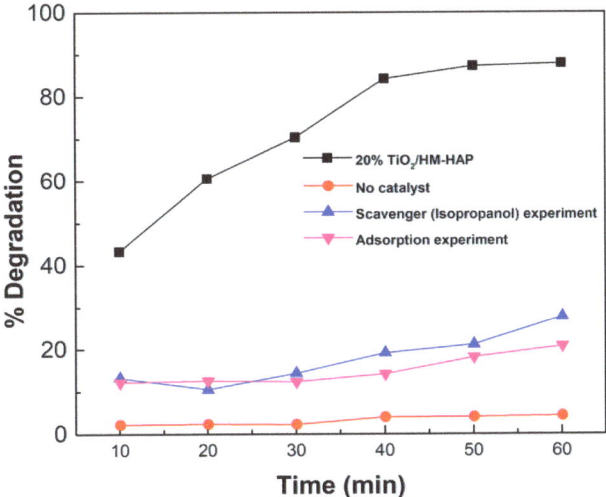

Figure 9. Plot representing the effect of a scavenger on the photolysis, adsorption, and photolysis experiment.

In addition, an examination was conducted on the impact of an OH scavenger on photocatalysis. Isopropanol was employed as an *OH scavenger for the experiment. The catalyst's catalytic efficiency decreases when a scavenger is present. This is illustrated using the graph presented in Figure 9. The predominant catalyst in this process is now clearly identified as the *OH radical [66].

In comparison with other recently researched sorbents, the adsorption capability of the Sa-modified hydroxyapatite as synthesized demonstrated promising findings. Zenefar et al. observed that the photodegradation potential of a HAp-TiO$_2$-ZnO photocatalyst for methylene blue (MB) and methyl orange (MO) dye was 95% and 45% after 2 h, respectively [67]. A Hap-TiO$_2$ nanocomposite demonstrated a degradation efficiency of 80% for methyl orange (5 mg/mL), as stated by Sharifat et al. [38]. Furthermore, Anmin et al. reported the degradation efficiency of titanium-substituted HAp for methylene blue (MB) to be 17%–37% under visible light and 39%–50% under UV light [68]. This literature review shows that the TiO$_2$-coated HM-HAP composite is an effective catalyst for the removal of dyes from water.

Combining the outcomes of SEM, DRS, FTIR, N$_2$ adsorption–desorption, and XRD analysis, the enhanced photocatalytic degradation performance of the TiO$_2$/HM-HAP coated composites in comparison with pure TiO$_2$ on the photodegradation of methyl red is associated with the creation of a uniform layer of TiO$_2$ particles on the highly porous structure of HM-HAP containing a large number of hydroxyl (OH) groups. These OH groups interact with the generated holes as Lewis bases, which results in the generation of hydroxyl radicals. The generated OH radicals will then oxidize the adsorbed molecules, hence improving the material's photocatalytic efficiency. Moreover, the hollow mesoporous structure of HM-HAP promotes the degradation performance of the composite by adsorbing the dye particles on its surface, as observed in Figure 7a. The relatively better catalytic performance of the 20% TiO$_2$/HM-HAP coated composites compared with those with a high concentration of TiO$_2$ is mostly determined by the existence of active sites, surface area, and availability of active substances.

3.4. Photocatalytic Degradation Mechanism of TiO$_2$/HM-HAP Composites on MR Dye

The possible mechanism of TiO$_2$/HM-HAP coated composites for the degradation of MR is shown in Figure 10. Generally, the UV irradiation of TiO$_2$-coated HM-HAP alters the electronic condition of the PO_4^{-3} group on the surface and generates a vacancy on HM-HAP. In the same way, when UV light is applied to TiO_2, electrons in the VB (valance band) move to the CB (conduction band), making an equal number of holes in the VB. The reaction is given as:

$$TiO_2 + h\nu \rightarrow e^- + h^+ \tag{3}$$

The reaction of a CB electron with O_2 produces superoxide radicals ($O_2^{\cdot -}$), which then oxidizes the organic compounds. The reaction is given as follows:

$$e^- + O_2 \rightarrow O_2^- \tag{4}$$

The VB hole interacts with the hydroxyl anions or the water to generate hydrogen peroxide.

$$h^+ + H_2O \rightarrow OH^\cdot + H^+ \tag{5}$$

$$h^+ + OH^- \rightarrow OH^\cdot \tag{6}$$

The hydrogen peroxide then breaks apart and releases hydroxyl radicals (OH·), which are powerful oxidizing agents that attack organic molecules that have adsorbed to the composite [37].

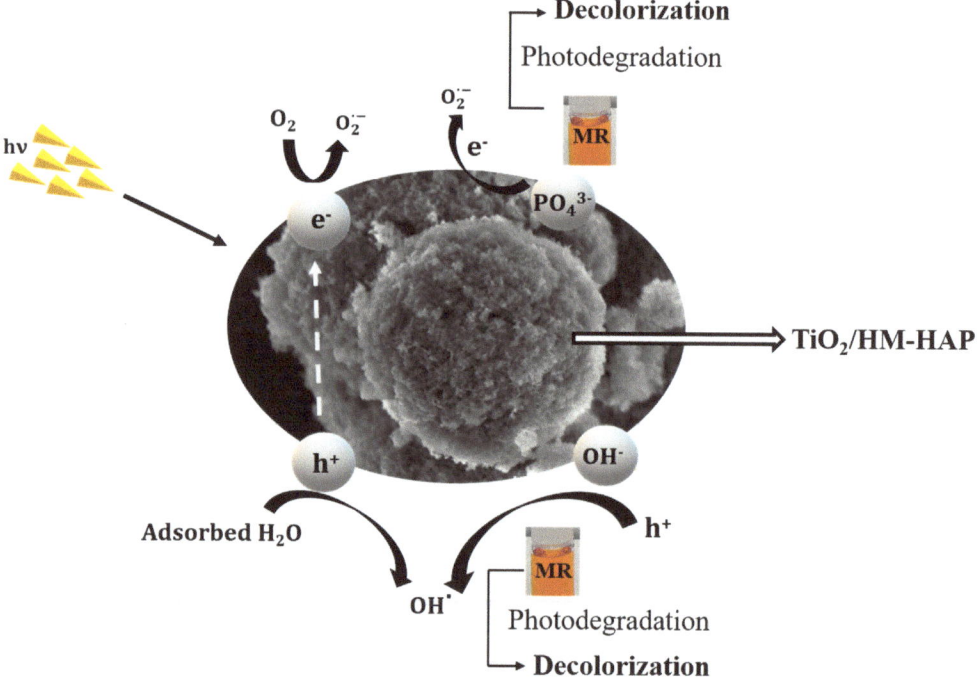

Figure 10. Possible mechanism of TiO_2/HM-HAP coated composites under UV irradiation for the photocatalytic degradation of MR.

4. Conclusions

The present research demonstrated the effective synthesis of TiO_2/HM-HAP coated composites by loading the TiO_2 at various concentrations, comprising 10%, 20%, 30%, 40%, and 50%, to enhance the degradation efficiency of hazardous dyes present in aqueous solutions and address water pollution issues. Several analysis approaches, including FTIR, XPS, XRD, SEM, and BET, were utilized to analyze the physical properties of all the synthesized TiO$_2$/HM-HAP coated composites. TiO$_2$ was evenly incorporated into the hollow mesoporous HAP particles, with an insignificant effect on the overall structure. The catalytic performance was examined via the degradation of MR in the presence of UV light. We observed that the MR removal ratio of the 20% TiO$_2$/HM-HAP coating was 88%, which was found to be the maximum among the HM-HAP, pure TiO$_2$, and TiO$_2$/HM-HAP coated composites. It was noted that the photocatalytic degradation process was compatible with the pseudo-first-order (PFO) kinetic model. Moreover, it was found that MR could be more easily degraded at pH 6 than in strongly acidic or alkaline environments. This work provides insight into the development of TiO$_2$/HM-HAP coated composites as potential materials for removing organic contaminants from wastewater. This will help with both water treatment and waste management.

Supplementary Materials: The following supporting information can be downloaded at https://www.mdpi.com/article/10.3390/coatings14080921/s1. It includes the details about band gap energy data and other supporting figures.

Author Contributions: Conceptualization, F.S. and W.Q.; Methodology, F.S.; Formal analysis, S.Y. and Y.P.; Investigation, S.Y.; Writing—original draft, F.S.; Project administration, W.Q. All authors have read and agreed to the published version of the manuscript.

Funding: This research was supported by the National Natural Science Foundation of China Grant No. 22378049.

Institutional Review Board Statement: Not applicable.

Informed Consent Statement: Not applicable.

Data Availability Statement: Data is contained within the article or Supplementary Materials.

Conflicts of Interest: The authors declare no conflicts of interest.

References

1. Badr, Y.; Abd El-Wahed, M.G.; Mahmoud, M.A. Photocatalytic degradation of methyl red dye by silica nanoparticles. *J. Hazard. Mater.* **2008**, *154*, 245–253. [CrossRef] [PubMed]
2. Zaheer, Z.; AL-Asfar, A.; Aazam, E.S. Adsorption of methyl red on biogenic Ag@Fe nanocomposite adsorbent: Isotherms, kinetics and mechanisms. *J. Mol. Liq.* **2019**, *283*, 287–298. [CrossRef]
3. Slokar, Y.M.; Majcen Le Marechal, A. Methods of decoloration of textile wastewaters. *Dye. Pigment.* **1998**, *37*, 335–356. [CrossRef]
4. Gökçen, F.; Özbelge, T.A. Pre-ozonation of aqueous azo dye (Acid Red-151) followed by activated sludge process. *Chem. Eng. J.* **2006**, *123*, 109–115. [CrossRef]
5. Khan, A.M.; Shafiq, F.; Khan, S.A.; Ali, S.; Ismail, B.; Hakeem, A.S.; Rahdar, A.; Nazar, M.F.; Sayed, M.; Khan, A.R. Surface modification of colloidal silica particles using cationic surfactant and the resulting adsorption of dyes. *J. Mol. Liq.* **2019**, *274*, 673–680. [CrossRef]
6. Ullah, F.; Ji, G.; Irfan, M.; Gao, Y.; Shafiq, F.; Sun, Y.; Ain, Q.U.; Li, A. Adsorption performance and mechanism of cationic and anionic dyes by KOH activated biochar derived from medical waste pyrolysis. *Environ. Pollut.* **2022**, *314*, 120271. [CrossRef] [PubMed]
7. Zhao, J.; Liu, H.; Xue, P.; Tian, S.; Sun, S.; Lv, X. Highly-efficient PVDF adsorptive membrane filtration based on chitosan@CNTs-COOH simultaneous removal of anionic and cationic dyes. *Carbohydr. Polym.* **2021**, *274*, 118664. [CrossRef] [PubMed]
8. Saulat, H.; Yang, J.; Yan, T.; Raza, W.; Song, W.; He, G. Tungsten incorporated mobil-type eleven zeolite membranes: Facile synthesis and tuneable wettability for highly efficient separation of oil/water mixtures. *Chin. J. Chem. Eng.* **2023**, *60*, 242–252. [CrossRef]
9. Badr, Y.; Mahmoud, M.A. Photocatalytic degradation of methyl orange by gold silver nano-core/silica nano-shell. *J. Phys. Chem. Solids* **2007**, *68*, 413–419. [CrossRef]
10. Kim, H.; Kim, T.; Gil Lee, D.; Weon Roh, S.; Lee, C. Nitrogen-centered radical-mediated C–H imidation of arenes and heteroarenes via visible light induced photocatalysis. *Chem. Commun.* **2014**, *50*, 9273–9276. [CrossRef] [PubMed]
11. Lucas, M.S.; Dias, A.A.; Sampaio, A.; Amaral, C.; Peres, J.A. Degradation of a textile reactive Azo dye by a combined chemical–biological process: Fenton's reagent-yeast. *Water Res.* **2007**, *41*, 1103–1109. [CrossRef] [PubMed]
12. Liu, Y.; Liu, C.Y.; Wei, J.H.; Xiong, R.; Pan, C.X.; Shi, J. Enhanced adsorption and visible-light-induced photocatalytic activity of hydroxyapatite modified Ag–TiO$_2$ powders. *Appl. Surf. Sci.* **2010**, *256*, 6390–6394. [CrossRef]
13. Natarajan, T.S.; Lee, J.Y.; Bajaj, H.C.; Jo, W.K.; Tayade, R.J. Synthesis of multiwall carbon nanotubes/TiO$_2$ nanotube composites with enhanced photocatalytic decomposition efficiency. *Catal. Today* **2017**, *282*, 13–23. [CrossRef]

14. Wu, H.; Yang, X.; Zhao, S.; Zhai, L.; Wang, G.; Zhang, B.; Qin, Y. Encapsulation of atomically dispersed Pt clusters in porous TiO_2 for semi-hydrogenation of phenylacetylene. *Chem. Commun.* **2022**, *58*, 1191–1194. [CrossRef] [PubMed]
15. Chen, S.; Fang, S.; Sun, Z.; Li, Z.; Wang, C.; Hu, Y.H. Thin-water-film-enhanced TiO_2-based catalyst for CO_2 hydrogenation to formic acid. *Chem. Commun.* **2022**, *58*, 787–790. [CrossRef] [PubMed]
16. Liu, H.; Lv, T.; Zhu, C.; Zhu, Z. Direct bandgap narrowing of TiO_2/MoO_3 heterostructure composites for enhanced solar-driven photocatalytic activity. *Sol. Energy Mater. Sol. Cells* **2016**, *153*, 1–8. [CrossRef]
17. Sukhadeve, G.K.; Bandewar, H.; Janbandhu, S.Y.; Jayaramaiah, J.R.; Gedam, R.S. Photocatalytic hydrogen production, dye degradation, and antimicrobial activity by Ag-Fe co-doped TiO_2 nanoparticles. *J. Mol. Liq.* **2023**, *369*, 120948. [CrossRef]
18. Ali, F.; Moin-ud-Din, G.; Iqbal, M.; Nazir, A.; Altaf, I.; Alwadai, N.; Siddiqua, U.H.; Younas, U.; Ali, A.; Kausar, A.; et al. Ag and Zn doped TiO_2 nano-catalyst synthesis via a facile green route and their catalytic activity for the remediation of dyes. *J. Mater. Res. Technol.* **2023**, *23*, 3626–3637. [CrossRef]
19. Liza, T.Z.; Tusher, M.M.H.; Anwar, F.; Monika, M.F.; Amin, K.F.; Asrafuzzaman, F.N.U. Effect of Ag-doping on morphology, structure, band gap and photocatalytic activity of bio-mediated TiO_2 nanoparticles. *Results Mater.* **2024**, *22*, 100559. [CrossRef]
20. Poudel, M.B.; Kim, A.A. Silver nanoparticles decorated TiO_2 nanoflakes for antibacterial properties. *Inorg. Chem. Commun.* **2023**, *152*, 110675. [CrossRef]
21. Cabrera-Rodríguez, O.; Trejo-Valdez, M.D.; Torres-SanMiguel, C.R.; Pérez-Hernández, N.; Bañuelos-Hernández, Á.; Manríquez-Ramírez, M.E.; Hernández-Benítez, J.A.; Rodríguez-Tovar, A.V. Evaluation of the performance of TiO_2 thin films doped with silver nanoparticles as a protective coating for metal prostheses. *Surf. Coatings Technol.* **2023**, *458*, 129349. [CrossRef]
22. Meroni, D.; Galloni, M.G.; Cionti, C.; Cerrato, G.; Falletta, E.; Bianchi, C.L. Efficient Day-and-Night NO_2 Abatement by Polyaniline/TiO_2 Nanocomposites. *Materials* **2023**, *16*, 1304. [CrossRef] [PubMed]
23. Pal, A.; Jana, T.K.; Chatterjee, K. Silica supported TiO_2 nanostructures for highly efficient photocatalytic application under visible light irradiation. *Mater. Res. Bull.* **2016**, *76*, 353–357. [CrossRef]
24. Eskandarian, M.R.; Fazli, M.; Rasoulifard, M.H.; Choi, H. Decomposition of organic chemicals by zeolite-TiO_2 nanocomposite supported onto low density polyethylene film under UV-LED powered by solar radiation. *Appl. Catal. B Environ.* **2016**, *183*, 407–416. [CrossRef]
25. Padmanabhan, S.K.; Pal, S.; Ul Haq, E.; Licciulli, A. Nanocrystalline TiO_2–diatomite composite catalysts: Effect of crystallization on the photocatalytic degradation of rhodamine B. *Appl. Catal. A Gen.* **2014**, *485*, 157–162. [CrossRef]
26. Zeng, G.; You, H.; Du, M.; Zhang, Y.; Ding, Y.; Xu, C.; Liu, B.; Chen, B.; Pan, X. Enhancement of photocatalytic activity of TiO_2 by immobilization on activated carbon for degradation of aquatic naphthalene under sunlight irradiation. *Chem. Eng. J.* **2021**, *412*, 128498. [CrossRef]
27. Sans, J.; Arnau, M.; Sanz, V.; Turon, P.; Alemán, C. Hydroxyapatite-based biphasic catalysts with plasticity properties and its potential in carbon dioxide fixation. *Chem. Eng. J.* **2022**, *433*, 133512. [CrossRef]
28. Sayed, I.R.; Farhan, A.M.; AlHammadi, A.A.; El-Sayed, M.I.; Abd El-Gaied, I.M.; El-Sherbeeny, A.M.; Al Zoubi, W.; Ko, Y.G.; Abukhadra, M.R. Synthesis of novel nanoporous zinc phosphate/hydroxyapatite nano-rods (ZPh/HPANRs) core/shell for enhanced adsorption of Ni^{2+} and Co^{2+} ions: Characterization and application. *J. Mol. Liq.* **2022**, *360*, 119527. [CrossRef]
29. Ding, Z.; Han, H.; Fan, Z.; Lu, H.; Sang, Y.; Yao, Y.; Cheng, Q.; Lu, Q.; Kaplan, D.L. Nanoscale Silk-Hydroxyapatite Hydrogels for Injectable Bone Biomaterials. *ACS Appl. Mater. Interfaces* **2017**, *9*, 16913–16921. [CrossRef] [PubMed]
30. Gong, M.; Liu, C.; Liu, C.; Wang, L.; Shafiq, F.; Liu, X.; Sun, G.; Song, Q.; Qiao, W. Biomimetic hydroxyapate/polydopamine composites with good biocompatibility and efficiency for uncontrolled bleeding. *J. Biomed. Mater. Res. Part B Appl. Biomater.* **2021**, *109*, 1876–1892. [CrossRef] [PubMed]
31. Verma, R.; Mishra, S.R.; Gadore, V.; Ahmaruzzaman, M. Hydroxyapatite-based composites: Excellent materials for environmental remediation and biomedical applications. *Adv. Colloid Interface Sci.* **2023**, *315*, 102890. [CrossRef] [PubMed]
32. Amenaghawon, A.N.; Anyalewechi, C.L.; Darmokoesoemo, H.; Kusuma, H.S. Hydroxyapatite-based adsorbents: Applications in sequestering heavy metals and dyes. *J. Environ. Manag.* **2022**, *302*, 113989. [CrossRef] [PubMed]
33. Zhang, J.; Yan, B.; Chen, T.; Tu, S.; Li, H.; Yang, Z.; Hao, T.; Chen, C. Piezoelectric hydroxyapatite synthesized from municipal solid waste incineration fly ash and its underlying mechanism for high efficiency in degradation of xanthate. *Chem. Eng. J.* **2024**, *493*, 152601. [CrossRef]
34. Kumar Yadav, M.; Hiren Shukla, R.; Prashanth, K.G. A comprehensive review on development of waste derived hydroxyapatite (HAp) for tissue engineering application. *Mater. Today Proc.* **2023**. [CrossRef]
35. Tong, H.; Shi, D.; Cai, H.; Liu, J.; Lv, M.; Gu, L.; Luo, L.; Wang, B. Novel hydroxyapatite (HAP)-assisted hydrothermal solidification of heavy metals in fly ash from MSW incineration: Effect of HAP liquid-precursor and HAP seed crystal derived from eggshell waste. *Fuel Process. Technol.* **2022**, *236*, 107400. [CrossRef]
36. Nishikawa, H. Surface changes and radical formation on hydroxyapatite by UV irradiation for inducing photocatalytic activation. *J. Mol. Catal. A Chem.* **2003**, *206*, 331–338. [CrossRef]
37. El, A.; Boumanchar, I.; Zbair, M.; Chhiti, Y.; Sahibed-Dine, A.; Bentiss, F.; Bensitel, M. The photocatalytic degradation of methylene bleu over TiO_2 catalysts supported on hydroxyapatite. *JMES* **2017**, *8*, 1301–1311. Available online: http://www.jmaterenvironsci.com/ (accessed on 7 October 2022).

38. Sharifat, S.; Zolgharnein, H.; Hamidifalahi, A.; Enayati-Jazi, M.; Hamid, E. Preparation and Characterization of HAp/TiO$_2$ Nanocomposite for Photocatalytic Degradation of Methyl Orange under UV-Irradiation. *Adv. Mater. Res.* **2014**, *829*, 594–599. [CrossRef]
39. Yao, J.; Zhang, Y.; Wang, Y.; Chen, M.; Huang, Y.; Cao, J.; Ho, W.; Lee, S.C. Enhanced photocatalytic removal of NO over titania/hydroxyapatite (TiO$_2$/HAp) composites with improved adsorption and charge mobility ability. *RSC Adv.* **2017**, *7*, 24683–24689. [CrossRef]
40. Narayan, R.B.; Goutham, R.; Srikanth, B.; Gopinath, K.P. A novel nano-sized calcium hydroxide catalyst prepared from clam shells for the photodegradation of methyl red dye. *J. Environ. Chem. Eng.* **2018**, *6*, 3640–3647. [CrossRef]
41. Shan, R.; Lu, L.; Gu, J.; Zhang, Y.; Yuan, H.; Chen, Y.; Luo, B. Photocatalytic degradation of methyl orange by Ag/TiO$_2$/biochar composite catalysts in aqueous solutions. *Mater. Sci. Semicond. Process.* **2020**, *114*, 105088. [CrossRef]
42. Xu, Q.; Feng, J.; Li, L.; Xiao, Q.; Wang, J. Hollow ZnFe$_2$O$_4$/TiO$_2$ composites: High-performance and recyclable visible-light photocatalyst. *J. Alloys Compd.* **2015**, *641*, 110–118. [CrossRef]
43. Kalaiarasi, S.; Jose, M. Dielectric functionalities of anatase phase titanium dioxide nanocrystals synthesized using water-soluble complexes. *Appl. Phys. A* **2017**, *123*, 512. [CrossRef]
44. Wei, J.; Shi, J.; Wu, Q.; Yang, L.; Cao, S. Hollow hydroxyapatite/polyelectrolyte hybrid microparticles with controllable size, wall thickness and drug delivery properties. *J. Mater. Chem. B* **2015**, *3*, 8162–8169. [CrossRef] [PubMed]
45. Zhu, X.; Shi, J.; Ma, H.; Chen, R.; Li, J.; Cao, S. Hierarchical hydroxyapatite/polyelectrolyte microcapsules capped with AuNRs for remotely triggered drug delivery. *Mater. Sci. Eng. C* **2019**, *99*, 1236–1245. [CrossRef] [PubMed]
46. Shafiq, F.; Liu, C.; Zhou, H.; Chen, H.; Yu, S.; Qiao, W. Adsorption mechanism and synthesis of adjustable hollow hydroxyapatite spheres for efficient wastewater cationic dyes adsorption. *Colloids Surfaces A Physicochem. Eng. Asp.* **2023**, *672*, 131713. [CrossRef]
47. Wang, S.; Zhou, S. Photodegradation of methyl orange by photocatalyst of CNTs/P-TiO$_2$ under UV and visible-light irradiation. *J. Hazard. Mater.* **2011**, *185*, 77–85. [CrossRef] [PubMed]
48. Xu, W.; Liu, B.; Wang, Y.; Xiao, G.; Chen, X.; Xu, W.; Lu, Y.P. A facile strategy for one-step hydrothermal preparation of porous hydroxyapatite microspheres with core–shell structure. *J. Mater. Res. Technol.* **2022**, *17*, 320–328. [CrossRef]
49. Shafiq, F.; Liu, C.; Zhou, H.; Chen, H.; Yu, S.; Qiao, W. Stearic acid-modified hollow hydroxyapatite particles with enhanced hydrophobicity for oil adsorption from oil spills. *Chemosphere* **2024**, *348*, 140651. [CrossRef] [PubMed]
50. Guo, Y.P.; Yao, Y.B.; Guo, Y.J.; Ning, C.Q. Hydrothermal fabrication of mesoporous carbonated hydroxyapatite microspheres for a drug delivery system. *Microporous Mesoporous Mater.* **2012**, *155*, 245–251. [CrossRef]
51. Li, W.; Tian, Y.; Li, H.; Zhao, C.; Zhang, B.; Zhang, H.; Geng, W.; Zhang, Q. Novel BiOCl/TiO$_2$ hierarchical composites: Synthesis, characterization and application on photocatalysis. *Appl. Catal. A Gen.* **2016**, *516*, 81–89. [CrossRef]
52. Abbasi, S.; Bayati, M.R.; Golestani-Fard, F.; Rezaei, H.R.; Zargar, H.R.; Samanipour, F.; Shoaei-Rad, V. Micro arc oxidized HAp–TiO$_2$ nanostructured hybrid layers-part I: Effect of voltage and growth time. *Appl. Surf. Sci.* **2011**, *257*, 5944–5949. [CrossRef]
53. Li, K.; Gao, S.; Wang, Q.; Xu, H.; Wang, Z.; Huang, B.; Dai, Y.; Lu, J. In-situ-reduced synthesis of Ti^{3+} self-doped TiO$_2$/g-C$_3$N$_4$ heterojunctions with high photocatalytic performance under LED light irradiation. *ACS Appl. Mater. Interfaces* **2015**, *7*, 9023–9030. [CrossRef] [PubMed]
54. Salarian, M.; Xu, W.Z.; Wang, Z.; Sham, T.K.; Charpentier, P.A. Hydroxyapatite-TiO$_2$-based Nanocomposites Synthesized in Supercritical CO$_2$ for Bone Tissue Engineering: Physical and Mechanical Properties. *ACS Appl. Mater. Interfaces* **2014**, *6*, 16918–16931. [CrossRef] [PubMed]
55. Anjaneyulu, U.; Priyadarshini, B.; Arul Xavier Stango, S.; Chellappa, M.; Geetha, M.; Vijayalakshmi, U. Preparation and characterisation of sol–gel-derived hydroxyapatite nanoparticles and its coatings on medical grade Ti-6Al-4V alloy for biomedical applications. *Mater. Technol.* **2017**, *32*, 800–814. [CrossRef]
56. Rath, P.C.; Singh, B.P.; Besra, L.; Bhattacharjee, S. Multiwalled Carbon Nanotubes Reinforced Hydroxyapatite-Chitosan Composite Coating on Ti Metal: Corrosion and Mechanical Properties. *J. Am. Ceram. Soc.* **2012**, *95*, 2725–2731. [CrossRef]
57. Goto, T.; Cho, S.H.; Ohtsuki, C.; Sekino, T. Selective adsorption of dyes on TiO$_2$-modified hydroxyapatite photocatalysts morphologically controlled by solvothermal synthesis. *J. Environ. Chem. Eng.* **2021**, *9*, 105738. [CrossRef]
58. Lin, Y.C.; Fang, Y.P.; Hung, C.F.; Yu, H.P.; Alalaiwe, A.; Wu, Z.Y.; Fang, J.Y. Multifunctional TiO$_2$/SBA-15 mesoporous silica hybrids loaded with organic sunscreens for skin application: The role in photoprotection and pollutant adsorption with reduced sunscreen permeation. *Colloids Surf. B Biointerfaces* **2021**, *202*, 111658. [CrossRef] [PubMed]
59. Singh, N.K.; Saha, S.; Pal, A. Methyl red degradation under UV illumination and catalytic action of commercial ZnO: A parametric study. *New Pub Balaban* **2014**, *56*, 1066–1076. [CrossRef]
60. Bouzid, T.; Grich, A.; Naboulsi, A.; Regti, A.; Alaoui Tahiri, A.; El Himri, M.; El Haddad, M. Adsorption of Methyl Red on porous activated carbon from agriculture waste: Characterization and response surface methodology optimization. *Inorg. Chem. Commun.* **2023**, *158*, 111544. [CrossRef]
61. Amari, A.; Yadav, V.K.; Pathan, S.K.; Singh, B.; Osman, H.; Choudhary, N.; Khedher, K.M.; Basnet, A. Remediation of Methyl Red Dye from Aqueous Solutions by Using Biosorbents Developed from Floral Waste. *Adsorpt. Sci. Technol.* **2023**, *2023*, 1532660. [CrossRef]

62. Waghmode, T.R.; Kurade, M.B.; Sapkal, R.T.; Bhosale, C.H.; Jeon, B.H.; Govindwar, S.P. Sequential photocatalysis and biological treatment for the enhanced degradation of the persistent azo dye methyl red. *J. Hazard. Mater.* **2019**, *371*, 115–122. [CrossRef] [PubMed]
63. Balu, P.; Asharani, I.V.; Thirumalai, D. Catalytic degradation of hazardous textile dyes by iron oxide nanoparticles prepared from Raphanus sativus leaves' extract: A greener approach. *J. Mater. Sci. Mater. Electron.* **2020**, *31*, 10669–10676. [CrossRef]
64. Ikram, M.; Naeem, M.; Zahoor, M.; Rahim, A.; Hanafiah, M.M.; Oyekanmi, A.A.; Shah, A.B.; Mahnashi, M.H.; Al Ali, A.; Jalal, N.A.; et al. Biodegradation of Azo Dye Methyl Red by Pseudomonas aeruginosa: Optimization of Process Conditions. *Int. J. Environ. Res. Public Health* **2022**, *19*, 9962. [CrossRef] [PubMed]
65. Bian, C.; Wang, Y.; Yi, Y.; Shao, S.; Sun, P.; Xiao, Y.; Wang, W.; Dong, X. Enhanced photocatalytic activity of S-doped graphitic carbon nitride hollow microspheres: Synergistic effect, high-concentration antibiotic elimination and antibacterial behavior. *J. Colloid Interface Sci.* **2023**, *643*, 256–266. [CrossRef]
66. Shahzad, W.; Badawi, A.K.; Rehan, Z.A.; Khan, A.M.; Khan, R.A.; Shah, F.; Ali, S.; Ismail, B. Enhanced visible light photocatalytic performance of $Sr_{0.3}(Ba,Mn)_{0.7}ZrO_3$ perovskites anchored on graphene oxide. *Ceram. Int.* **2022**, *48*, 24979–24988. [CrossRef]
67. Yeasmin, Z.; Alim, A.; Ahmed, S.; Rahman, M.M.; Masum, S.M.; Ghosh, A.K. Synthesis, Characterization and Efficiency of $HAp-TiO_2-ZnO$ Composite as a Promising Photocatalytic Material. *Trans. Indian Ceram. Soc.* **2018**, *77*, 161–168. [CrossRef]
68. Hu, A.; Li, M.; Chang, C.; Mao, D. Preparation and characterization of a titanium-substituted hydroxyapatite photocatalyst. *J. Mol. Catal. A Chem.* **2007**, *267*, 79–85. [CrossRef]

Disclaimer/Publisher's Note: The statements, opinions and data contained in all publications are solely those of the individual author(s) and contributor(s) and not of MDPI and/or the editor(s). MDPI and/or the editor(s) disclaim responsibility for any injury to people or property resulting from any ideas, methods, instructions or products referred to in the content.

Article

Experimental and Adsorption Kinetics Study of Hg^0 Removal from Flue Gas by Silver-Loaded Rice Husk Gasification Char

Ru Yang [1,2,*], Yongfa Diao [2], Hongbin Liu [1] and Yihang Lu [1]

[1] Nanxun Innovation Institute, Zhejiang University of Water Resources and Electric Power, Hangzhou 310018, China
[2] College of Environmental Science and Engineering, Donghua University, Shanghai 201620, China
* Correspondence: yangru@zjweu.edu.cn

Abstract: Coal holds a significant position in China's energy consumption structure. However, the release of Hg^0 during coal combustion poses a serious threat to human health. Traditional activated carbon for Hg^0 removal is expensive; finding efficient, inexpensive and renewable adsorbents for Hg^0 removal has become a top priority. Rice husk gasification char (RHGC) is a solid waste generated by biomass gasification power generation, which, loaded with silver to remove Hg^0, could achieve the purpose of waste treatment. This paper examines the Hg^0 removal performance of silver-loaded rice husk gasification char (SRHGC) under different operating conditions through experimental analysis. This study employed quasi-first-order, quasi-second-order, and internal diffusion kinetics adsorption equations to model the amount of Hg^0 removed by SRHGC at different temperatures, thereby inferring the reaction mechanism. The results indicate that Hg^0 removal efficiency of SRHGC increased by about 80%. The Hg^0 removal ability was directly related to silver load, and the amount of Hg^0 removed by SRHGC did not a exhibit a simple inverse relationship with particle size. Additionally, the Hg^0 removal efficiency of SRHGC declined with increasing adsorption temperature. The removal of Hg^0 by SRHGC conformed to the quasi-second-order kinetic equation, with the adsorption rate constant decreasing as the temperature rose, consistent with experimental observations. This paper provides both experimental and theoretical references for future modification and optimization of RHGC for coal-fired flue gas treatment, and also offers valuable insights into Hg^0 removal by carbon-based adsorbents.

Keywords: rice husk gasification char; silver-loaded; Hg^0; adsorption kinetics; mechanism

Citation: Yang, R.; Diao, Y.; Liu, H.; Lu, Y. Experimental and Adsorption Kinetics Study of Hg^0 Removal from Flue Gas by Silver-Loaded Rice Husk Gasification Char. *Coatings* **2024**, *14*, 797. https://doi.org/10.3390/coatings14070797

Academic Editor: Eduardo Guzmán

Received: 22 May 2024
Revised: 20 June 2024
Accepted: 24 June 2024
Published: 26 June 2024

Copyright: © 2024 by the authors. Licensee MDPI, Basel, Switzerland. This article is an open access article distributed under the terms and conditions of the Creative Commons Attribution (CC BY) license (https://creativecommons.org/licenses/by/4.0/).

1. Introduction

With the development of the economy and the improvement of living standards, the demand for energy and electricity has grown rapidly [1]. In 2022, global energy consumption reached an all-time high, with primary energy consumption totaling 20.6 billion tons of standard coal, marking a 2.2% year-on-year increase and essentially returning to the average growth level prior to the pandemic [2]. Coal would occupy a long-term position in China's energy structure. In 2023, China consumed 5.72 billion tons of standard coal, accounting for 55.3% of its total energy consumption. The flue gas released by coal burning contains CO_2, NO_X, SO_2, and the heavy metal mercury; these pollutants can cause serious environmental pollution. CO_2 causes the greenhouse effect, SO_2 and NO_X cause acid rain, and mercury causes mental illness [3,4]. The control technologies for SO_2, NO_X, and CO_2 are relatively mature and could effectively remove these pollutants using adsorbents or devices [5,6]. However, Hg^0 removal technology is still in the development stage. Hg^0 is highly toxic, volatile, lipophilic, and bio-accumulative, and would seriously harm human health [7–10]. Furthermore, Hg^0 is insoluble in water, making it difficult to remove using traditional separation technologies [11].

Activated carbon (AC) is effective in removing Hg^0 from coal-fired flue gas, but the cost is high [12–14]. Therefore, finding an efficient and cost-effective adsorbent is imperative. Rice husk gasification char (RHGC), a solid waste from biomass gasification power generation, has gained increasing attention for recycling due to the "zero-waste city" concept [15]. RHGC possesses highly active surface functional groups and a developed microporous structure, making it an effective adsorbent for Hg^0 in flue gas. Loading RHGC with silver (SRHGC) enhances its Hg^0 removal capabilities and supports waste treatment objectives. Additionally, SRHGC can be regenerated after heating, making it a potential long-term adsorbent [16,17]. There are limited studies on Hg^0 removal using RHGC, especially after silver loading. This article provides valuable insights and a reference for future researchers studying Hg^0 removal with RHGC.

Removing Hg^0 with adsorbents is a complex process involving both surface adsorption and internal diffusion. Domestic and international scholars have predicted the performance of carbon-based adsorbents for Hg^0 removal by establishing various models [18]. This study investigated the Hg^0 removal performance of SRHGC under different conditions through experimental studies. Additionally, the quasi-first-order kinetic equation, the quasi-second-order kinetic equation, and the intraparticle diffusion equation were used to simulate the Hg^0 removal behavior of SRHGC at different adsorption temperatures, providing further insights into the Hg^0 removal mechanism. In the context of carbon neutrality and carbon peak goals, utilizing inexpensive and renewable SRHGC for Hg^0 removal is particularly relevant. Applying simulated kinetics methods to predict experimental results enhances the efficiency, economy, and predictability of the experiments, offering significant economic and practical benefits.

2. Materials and Methods

In the experiment, RHGC was obtained from the Jiangsu Gaoyou biomass gasification power plant. The carbon content was 43.58%, determined using a German Elementar Vario ELIII element analyzer (analytical precision C $\leq$ 0.1 abs). After drying, grinding, and sieving, particle sizes of 97–125 µm, 125–200 µm, and 200–450 µm were selected for use. RHGC (particle size 125–200 µm) was impregnated with 20% hydrochloric acid for 30 min, placed in a vacuum oven at 50 °C for 2 h, cooled to room temperature, washed to neutrality with deionized water, and dried for later use. A total of 6 mL of silver nitrate solution with a mass concentration of 2 mg/mL was placed in a beaker, and the pH was adjusted to 9.7 with ammonia water. A total of 300 mg of dried RHGC was added and the mixture was shaken at a constant temperature of 298 K for 24 h. RHGC was then filtered and placed in an electric furnace tube, where $Ag(NH_3)^{2+}$ was reduced to elemental silver under a N_2 atmosphere at 120 °C for 4 h. After the reduction process was complete, heating was stopped and the sample, which was silver-loaded rice husk gasification char (labeled as SRHGC, mass ratio of $AgNO_3$ to RHGC of approximately 40 mg/g), was allowed to cool to room temperature before being removed and stored in a desiccator for later use. The same method was used to prepare SRHGC-20 (mass ratio of $AgNO_3$ to RHGC was about 20 mg/g) and SRHGC-60 (mass ratio of $AgNO_3$ to RHGC was about 60 mg/g), which were stored in a desiccator for later use. The experiment was conducted in a fixed-bed experimental setup [19].

3. Results

3.1. X-ray Diffraction

Qualitative analysis of the phase and composition of SRHGC was conducted using a Rigaku D/max-2550 PC XRD analyzer from Japan. The XRD pattern of SRHGC is shown in Figure 1. A relatively sharp diffraction peak appeared near the diffraction angle $2\theta = 24°$, indicating that SRHGC was amorphous with a certain degree of crystallinity. A very sharp diffraction peak at $2\theta = 38°$ was observed, which was a typical characteristic diffraction peak of elemental silver, confirming the presence of elemental silver in RHGC. The height and sharpness of this peak reflected the silver content and crystallinity in the sample. Additionally, a sharp diffraction peak at $2\theta = 78°$ was noted, caused by silver entering RHGC through ion exchange, further confirming the successful loading of silver onto RHGC.

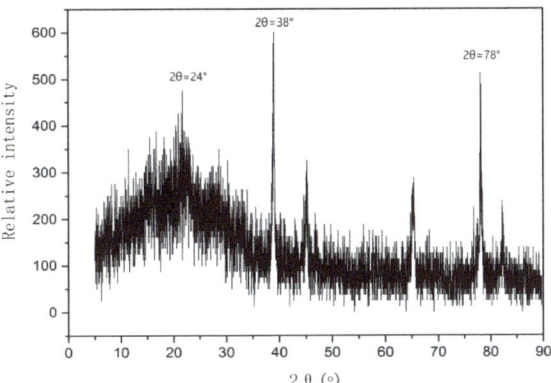

Figure 1. XRD pattern of SRHGC.

3.2. Hg^0 Removal Performance of SRHGC under Different Conditions

3.2.1. Hg^0 Removal Performance of Different Adsorbents

For the experiment, 300 mg each of RHGC, SRHGC, and AC were selected, along with an additional 450 mg of SRHGC (labeled as 2-SRHGC). The experimental conditions were set as follows: the temperature was maintained at 160 °C, the equilibrium gas was N_2 with a flow rate of 3 L/min, the inlet concentration of Hg^0 was 38.6 µg/m³, and the adsorption time was 120 min. The experimental results are shown in Figure 2.

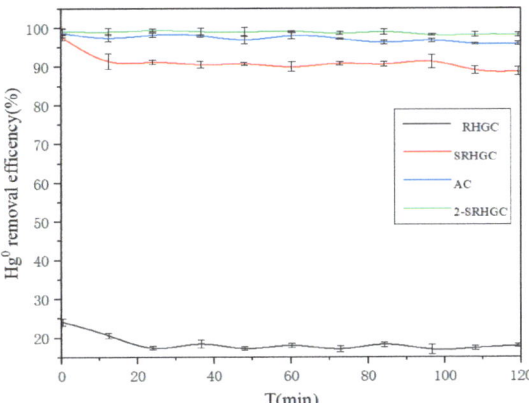

Figure 2. Hg^0 removal performance of different adsorbents.

Figure 2 shows that the Hg^0 removal efficiency stabilized after 20 min. The Hg^0 removal efficiencies of the four adsorbents followed this order: RHGC < SRHGC < AC < 2-SRHGC. The Hg^0 removal efficiency of RHGC increased by nearly 80%. Although the Hg^0 removal efficiency of an equivalent amount of SRHGC was lower than that of AC, increasing the quantity of SRHGC to 450 mg resulted in a higher Hg^0 removal efficiency compared to AC. This improvement was attributed to the increased mass, which provided a longer contact time between the adsorbent and mercury vapor, thereby enhancing Hg^0 removal efficacy. As a cost-effective industrial solid waste, SRHGC has the potential to replace AC for Hg^0 removal.

3.2.2. Effect of Different Silver Loads on Hg^0 Removal Performance of SRHGC

For the experiment, 300 mg each of SRHGC-20, SRHGC, and SRHGC-60 were selected. The experimental conditions were as follows: the temperature was maintained at 160 °C, the equilibrium gas was N_2 with a flow rate of 3 L/min, the inlet concentration of Hg^0 was 38.6 μg/m^3, and the adsorption time was 120 min. The experimental results are shown in Figure 3.

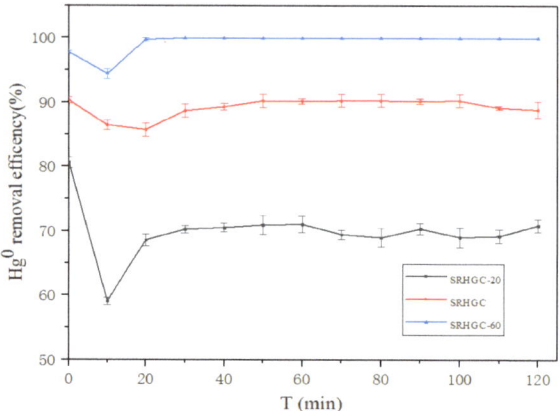

Figure 3. Effect of different silver loads on Hg^0 removal performance of SRHGC.

As seen in Figure 3, SRHGC-60 exhibited the highest Hg^0 removal efficiency, followed by SRHGC and then SRHGC-20. The Hg^0 removal ability was directly related to the silver load. After adsorption stabilization, the Hg^0 removal efficiency of SRHGC-60 reached 100%, SRHGC achieved about 90%, and SRHGC-20 attained 70%. Considering both silver load and efficiency, SRHGC demonstrated the best overall performance in removing Hg^0.

3.2.3. Effect of Different Particle Sizes on Hg^0 Removal Performance of SRHGC

For the experiment, 300 mg of SRHGC with particle sizes of 97–125 μm, 125–200 μm, and 200–450 μm were selected. The experimental conditions were as follows: the temperature was maintained at 160 °C, the equilibrium gas was N_2 with a flow rate of 3 L/min, the inlet concentration of Hg^0 was 38.6 μg/m^3, and the adsorption time was 120 min. The experimental results are shown in Figure 4.

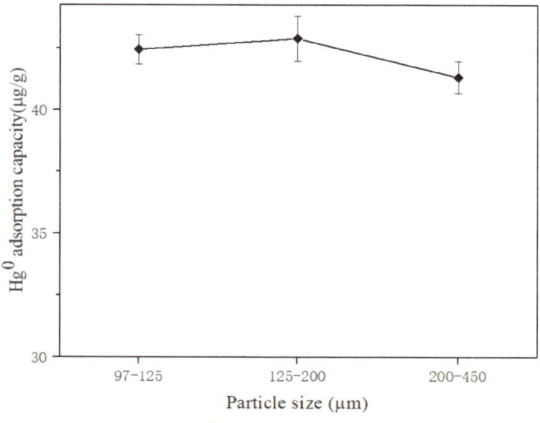

Figure 4. Effect of different particle sizes on Hg^0 removal performance of SRHGC.

As shown in Figure 4, the Hg^0 adsorption capacity was 41.79 μg/g for particles sized 200–450 μm, increased to 43.54 μg/g for particles sized 125–200 μm, and was 42.87 μg/g for particles sized 97–125 μm. This variation was due to the relationship between particle size and factors such as mass transfer resistance, specific surface area, and the internal diffusion coefficient of SRHGC. When particle size decreased, the specific surface area increased, and the internal diffusion coefficient also rose, facilitating the diffusion of Hg^0 to the surface of the adsorption layer and enhancing the adsorption of Hg^0 by SRHGC. At this stage, the positive effects dominated. However, further reduction in particle size increased the mass transfer resistance of the adsorption layer, leading to a rise in penetration pressure drop. Beyond a certain range, the specific surface area of SRHGC no longer increased with decreasing particle size, and the pressure drop continued to rise, weakening the effect of physical adsorption. Consequently, the negative effects became more significant. Therefore, to improve the Hg^0 adsorption performance of the adsorbent, the positive effects should outweigh the negative effects. Based on these findings, a particle size of 125–200 μm was selected for this experiment [20,21].

3.2.4. Effect of Different Temperatures on Hg^0 Removal Performance of SRHGC

For this experiment, 300 mg of SRHGC with a particle size of 125–200 μm was selected. The experimental temperatures were set at 120 °C, 160 °C, and 200 °C. The equilibrium gas was N^2 with a flow rate of 3 L/min, the inlet concentration of Hg^0 was 38.6 μg/m^3, and the adsorption time was 120 min. The results are shown in Figure 5. The Hg^0 adsorption capacities of SRHGC at 120 °C, 160 °C, and 200 °C were approximately 44.09 μg/g, 43.54 μg/g, and 41.15 μg/g, respectively. As the temperature increased, the Hg^0 adsorption capacity consistently decreased. This trend could be attributed to the nature of adsorption pores on the surface of SRHGC, where physical adsorption was predominant, and gaseous mercury was preferentially adsorbed at lower temperatures. As the temperature rose, the chemisorption rate surpassed the physical adsorption rate, potentially destroying the chemical bonds of oxygen-containing functional groups. Moreover, Hg^0 might degrade on the surface of SRHGC due to high temperatures, leading to a decrease in chemisorption capacity. Consequently, the adsorption capacity of SRHGC diminished at higher temperatures, with the rate of decline accelerating with temperature. Based on these findings, 160 °C was selected for this experiment.

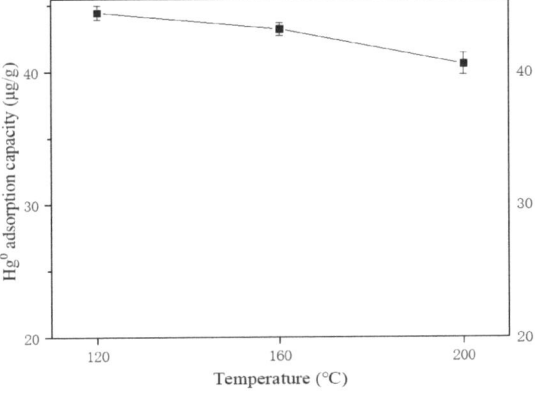

Figure 5. Effect of different temperatures on Hg^0 removal performance of SRHGC.

3.3. Adsorption Kinetics Study of Hg^0 Removal by SRHGC

The experimental data for Hg^0 removal by SRHGC at 120 °C (SRHCC-120), 160 °C (SRHCC-160), and 200 °C (SRHCC-200) were analyzed and fitted by using the quasi-first-order kinetic equation [22], the quasi-second-order kinetic equation [23] and the in-particle diffusion equation [24], the aim of which was to further study the mechanism of removing

Hg⁰ from SRHGC at different temperatures. In the figure below, q_e is the adsorption amount of mercury on the adsorbent at adsorption equilibrium (μg/g); q_t is the adsorption amount of mercury on the adsorbent at time t (μg/g); t is the adsorption time (min); k_1 is the rate constant of the adsorption quasi-first-order model (min^{-1}); k_2 is the rate constant of the bi-media rate equation (g/μg·min); C is related to the thickness at the boundary (μg/g); h and R^2 are the fitting parameters. The results are shown below.

3.3.1. Quasi-First-Order Kinetic Equation Fitting

According to the kinetic fitting results in Figure 6, the relevant kinetic parameters and correlation coefficients were obtained, as shown in Table 1.

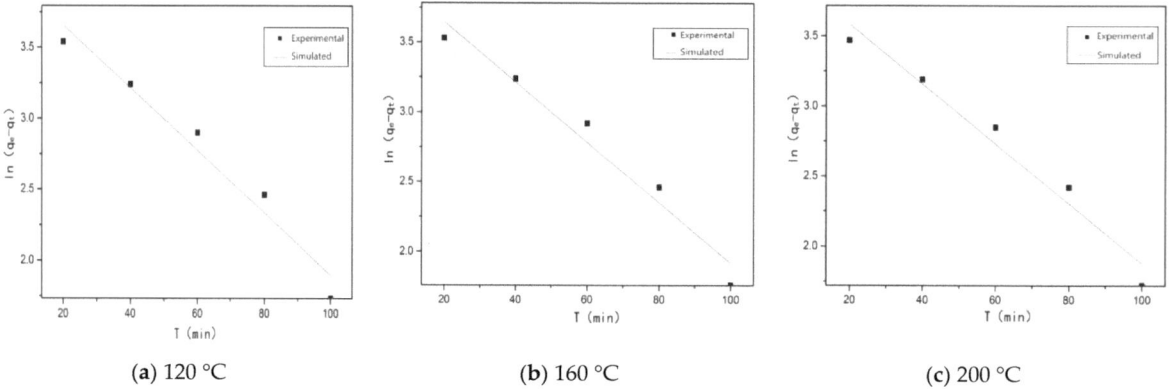

(a) 120 °C (b) 160 °C (c) 200 °C

Figure 6. Quasi-first-order kinetic simulation of SRHGC Hg⁰ removal.

Table 1. Related parameters of quasi-first-order dynamic equation.

Sample	q_e	K_1	R^2
SRHGC-120	44.09	0.022	0.95199
SRHGC-160	43.54	0.0216	0.95164
SRHGC-200	41.15	0.02135	0.95307

3.3.2. Quasi-Second-Order Kinetic Equation Fitting

According to the kinetic fitting results in Figure 7, the relevant kinetic parameters and correlation coefficients were obtained, as shown in Table 2.

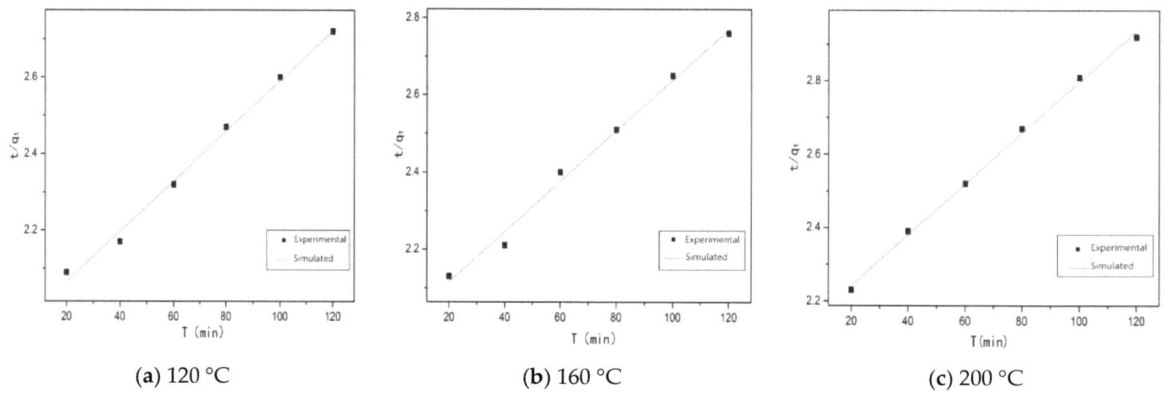

(a) 120 °C (b) 160 °C (c) 200 °C

Figure 7. Quasi-secondary dynamics simulation of SRHGC Hg⁰ removal.

Table 2. Related parameters of pseudo-second-order kinetic equation.

Sample	h	K_2	R^2
SRHGC-120	0.517	2.22×10^{-5}	0.99348
SRHGC-160	0.504	2.15×10^{-5}	0.99623
SRHGC-200	0.475	2.11×10^{-5}	0.99713

3.3.3. In-Particle Diffusion Equation Fitting

According to the kinetic fitting results in Figure 8, the relevant kinetic parameters and correlation coefficients were obtained, as shown in Table 3.

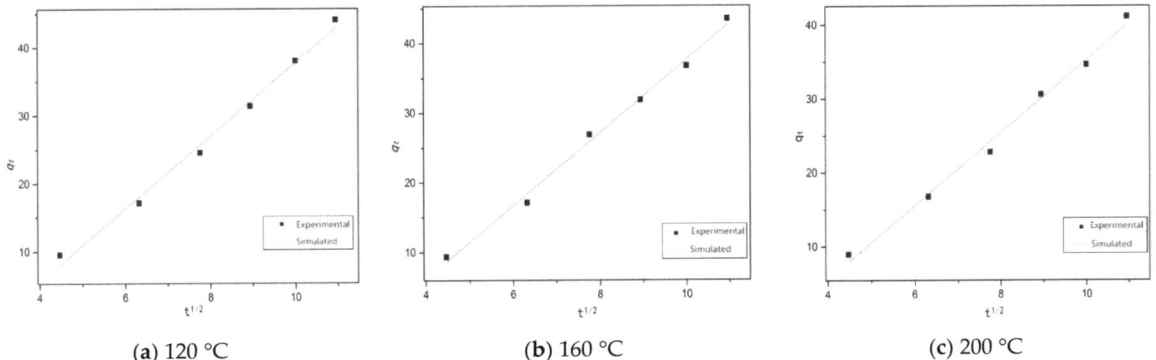

(a) 120 °C　　(b) 160 °C　　(c) 200 °C

Figure 8. Kinetic simulation of internal diffusion of SRHGC Hg^0 removal.

Table 3. Intraparticle diffusion equation-related parameters.

Sample	C	K_1	R^2
SRHGC-120	−15.85088	5.36538	0.99063
SRHGC-160	−14.69875	5.23916	0.99233
SRHGC-200	−14.05459	4.938	0.99103

Figures 6–8 show the fitting of Hg^0 adsorption capacity of SRHGC at different temperatures using the quasi-first-order, quasi-second-order, and internal diffusion kinetic equations, respectively. Tables 1–3 present the parameters obtained by fitting three kinetic equations at different temperatures. The fact that none of the lines in Figure 4 pass through the origin indicates that intraparticle diffusion is not the sole factor controlling the adsorption process. The adsorption of Hg^0 on the adsorbent surface was divided into two stages: internal diffusion and surface adsorption. In the initial stage, the adsorption rate was faster, while the internal diffusion rate was slower. The R^2 of the quasi-second-order kinetic fitting parameters were all above 0.993. The fitting curves were also very consistent with the experimental data of SRHGC, indicating that the adsorption of Hg^0 on SRHGC conforms to the quasi-second-order kinetic equation. It can be seen from Tables 1–3 that the adsorption rate constant of the quasi-first-order, quasi-second-order, and internal diffusion kinetic equations decreases with the increase in temperature, indicating that the adsorption performance of the adsorbent Hg^0 decreases with the increase in temperature, hindering the adsorption of Hg^0, which was consistent with the experimental results.

4. Conclusions

This paper investigated the Hg^0 removal performance of silver-loaded rice husk gasification char (SRHGC) under various operating conditions through experimental studies. The adsorption kinetics were analyzed using quasi-first-order, quasi-second-order, and intraparticle diffusion equations to model the amount of Hg^0 adsorbed by SRHGC at different temperatures, thereby inferring the reaction mechanism. The results show that the Hg^0 removal efficiency of SRHGC improved by approximately 80%. The amount of Hg^0 removed by SRHGC did not exhibit a simple inverse relationship with particle size, and the Hg^0 removal efficiency decreased with increasing adsorption temperature. The process of removing Hg^0 from SRHGC followed the quasi-second-order kinetic equation, with the adsorption rate constant decreasing as the temperature increased, consistent with the experimental findings. This study provides both experimental and theoretical references for the subsequent modification and selection of adsorbents. It also lays a solid foundation for the application of RHGC in the environmental protection field, particularly in the efficient removal of Hg^0 from flue gas, offering significant economic and practical benefits.

Author Contributions: Survey, R.Y. and Y.D.; experiment, R.Y.; data analysis, R.Y.; draw, Y.L.; writing—original draft preparation, R.Y.; writing—review and editing, R.Y. and H.L. All authors have read and agreed to the published version of the manuscript.

Funding: This work is supported by "Nanxun Scholars program of ZJWEU" (RC2022010919).

Institutional Review Board Statement: Not applicable.

Informed Consent Statement: Not applicable.

Data Availability Statement: Data are contained within the article.

Conflicts of Interest: The authors declare no conflicts of interest.

References

1. Liu, H.; Sun, Q.; Zhang, H.; Cheng, J.; Li, Y.; Zeng, Z.; Zhang, S.; Xu, X.; Ji, F.; Li, D.; et al. The application road of silicon-based anode in lithium-ion batteries: From liquid electrolyte to solid-state electrolyte. *Energy Storage Mater.* **2023**, *55*, 244–263. [CrossRef]
2. Xie, K. Challenges and approaches of clean and efficient coal conversion in the context of new energy system development. *J. Coal Sci.* **2024**, *1*, 1–13. [CrossRef]
3. Aaron, D.; Tsouris, C. Separation of CO2 from Flue Gas: A review. *Sep. Sci. Technol.* **2005**, *40*, 321–348. [CrossRef]
4. Zhong, L.; Xiao, P.; Jiang, J.; Guo, T.; Guo, A. Coal-fired power plants of atmospheric mercury emission monitoring method analysis and test study. *Proc. CSEE* **2012**, *32*, 158–163. [CrossRef]
5. Hassanpouryouzband, A.; Yang, J.; Tohidi, B.; Chuvilin, E.M.; Istomin, V.; Bukhanov, B.A.; Cheremisin, A. CO2 Capture by Injection of Flue Gas or CO2–N2 Mixtures into Hydrate Reservoirs: Dependence of CO2 Capture Efficiency on Gas Hydrate Reservoir Conditions. *Environ. Sci. Technol.* **2018**, *52*, 4324–4330. [CrossRef] [PubMed]
6. Hassanpouryouzband, A.; Joonaki, E.; Farahani, M.V.; Takeya, S.; Ruppel, C.; Yang, J.; English, N.J.; Schicks, J.M.; Edlmann, K.; Mehrabian, H. Gas hydrates in sustainable chemistry. *Chem. Soc. Rev.* **2020**, *49*, 5225–5309. [CrossRef] [PubMed]
7. Draseh, G.; O'Reilly, S.B.; Beinhoff, C.; Roider, G.; Maydl, S. The Mt. Diwata Study on the Philippines 1999 Assessing Mereury Intoxication of the Population by Small Seale Gold Mining. *Sci. Total Environ.* **2001**, *267*, 151–168. [CrossRef]
8. Zeng, Z.F.; Liu, T.X.; Li, H.M.; Hao, R.L. Research progress of adsorbents for mercury removal from coal flue gas. *Appl. Chem. Ind.* **2023**, *52*, 1170–1174. [CrossRef]
9. Abbasi, N.M.; Hameed, M.U.; Nasim, N.; Ahmed, F.; Altaf, F.; Shahida, S.; Fayyaz, S.; Sabir, S.M.; Bocchetta, P. Plasmonic Nano Silver: An Efficient Colorimetric Sensor for the Selective Detection of Hg2+ Ions in Real Samples. *Coatings* **2022**, *12*, 763. [CrossRef]
10. Li, Y.; Yu, J.; Liu, Y.; Huang, R.; Wang, Z.; Zhao, Y. A review on removal of mercury from flue gas utilizing existing air pollutant control devices (apcds). *J. Hazard. Mater.* **2022**, *427*, 128132. [CrossRef]
11. Zhang, Y.; Zeng, L.; Xu, L.; Liu, X.; Li, L.; Wang, H. Removal of Elemental Mercury from Simulated Flue Gas by a Copper-Based ZSM-5 Molecular Sieve. *Coatings* **2022**, *12*, 772. [CrossRef]
12. Wu, J.; Zhao, Z.; Huang, T.; Sheng, P.; Zhang, J.; Tian, H.; Zhao, X.; Zhao, L.; He, P.; Ren, J.; et al. Removal of elemental mercury by Ce-Mn co-modified activated carbon catalyst. *Catal. Commun.* **2017**, *93*, 62–66. [CrossRef]
13. Rungnim, C.; Promarak, V.; Hannongbua, S.; Kungwan, N.; Namuangruk, S. Complete reaction mechanisms of mercury oxidation on halogenated activated carbon. *J. Hazard. Mater.* **2016**, *310*, 253–260. [CrossRef] [PubMed]
14. Sun, P.; Zhang, B.; Zeng, X.; Luo, G.; Li, X.; Yao, H.; Zheng, C. Deep study on effects of activated carbon's oxygen functional groups for elemental mercury adsorption using temperature programmed desorption method. *Fuel* **2017**, *200*, 100–106. [CrossRef]
15. Li, Y.; Zhang, Y.; Zhang, S. A high proportion reuse of RAP in plant-mixed cold recycling technology and its benefits analysis. *Coatings* **2022**, *12*, 1283. [CrossRef]
16. Sun, H. Study on Controllable Synthesis of Molecular Sieve Loaded Silver Composite and Performance in Removing Hg0 from Natural Gas. Master's Thesis, Shandong University of Science and Technology, Qingdao, China, 2021.

17. Xu, H.; Qu, Z.; Huang, W.; Mei, J.; Chen, W.; Zhao, S.; Yan, N. Regenerable Ag/graphene sorbent for elemental mercury capture at ambient temperature. *Colloids Surf. A Physicochem. Eng. Asp.* **2015**, *476*, 47683–47689. [CrossRef]
18. Lee, T.G. Study of Mercury Kinetics and Control Methodogies in Simulates Combustion Flue Gases. Bachelor's Thesis, University of Cincinnati, Cincinnati, OH, USA, 1999.
19. Yang, R.; Diao, Y.; Befkadu, A. Removal of Hg0 from simulated flue gas over silver-loaded rice husk gasification char. *R. Soc. Open Sci.* **2018**, *5*, 180248. [CrossRef] [PubMed]
20. Joseph, R.F.; Radisav, D.V.; Wei, L.; Robert, C.T. Modeling Powdered Activated Carbon Injection for the Uptake of Elemental Mercury Vapors. *J. Air Waste Manag. Assoc.* **1998**, *48*, 1051–1059. [CrossRef]
21. Yan, R.; Liang, D.T.; Tsen, L.; Wong, Y.P.; Lee, Y.K. Bench-scale experimental evaluation of carbon performance on mercury vapour adsorption. *Fuel* **2004**, *83*, 2401–2409. [CrossRef]
22. Skodras, G.; Diamantopoulou, I.; Pantoleontos, G.; Sakellaropoulos, G.P. Kinetic studies of elemental mercury adsorption in activated carbon fixed bed reactor. *J. Hazard. Mater.* **2008**, *158*, 1–13. [CrossRef]
23. Ho, Y.S.; Mckay, G. Pseudo-second order model for sorption processes. *Process Biochem.* **1999**, *34*, 451–465. [CrossRef]
24. Zhou, Q. Experimental and Mechanism Study of Mercury Removal by Injection of Modified Adsorbent. Doctoral Thesis, Southeast University, Nanjing, China, 2016.

Disclaimer/Publisher's Note: The statements, opinions and data contained in all publications are solely those of the individual author(s) and contributor(s) and not of MDPI and/or the editor(s). MDPI and/or the editor(s) disclaim responsibility for any injury to people or property resulting from any ideas, methods, instructions or products referred to in the content.

MDPI AG
Grosspeteranlage 5
4052 Basel
Switzerland
Tel.: +41 61 683 77 34

Coatings Editorial Office
E-mail: coatings@mdpi.com
www.mdpi.com/journal/coatings

Disclaimer/Publisher's Note: The statements, opinions and data contained in all publications are solely those of the individual author(s) and contributor(s) and not of MDPI and/or the editor(s). MDPI and/or the editor(s) disclaim responsibility for any injury to people or property resulting from any ideas, methods, instructions or products referred to in the content.

www.ingramcontent.com/pod-product-compliance
Lightning Source LLC
LaVergne TN
LVHW070000100526
838202LV00019B/2594